Introduction
to Biomedical Engineering

Introduction to Biomedical Engineering

Michael M. Domach

Carnegie Mellon University

Upper Saddle River, New Jersey 07458

Library of Congress Cataloging-in-Publication Data

Domach, Michael M.
    Introduction to biomedical engineering / Michael M. Domach.
        p. cm.
    Includes bibliographical references and index.
    ISBN 0-13-061977-9
    1. Biomedical engineering. I. Title.

R856.D665 2004
610'.28--dc22

2003066434

Vice President and Editorial Director, ECS: *Marcia J. Horton*
Acquisitions Editor: *Dorothy Marrero*
Vice President and Director of Production and Manufacturing, ESM: *David W. Riccardi*
Executive Managing Editor: *Vince O'Brien*
Managing Editor: *David A. George*
Production Editor: *Daniel Sandin*
Director of Creative Services: *Paul Belfanti*
Art Director: *Jayne Conte*
Cover Designer: *Bruce Kenselaar*
Art Editor: *Greg Dulles*
Manufacturing Manager: *Trudy Pisciotti*
Manufacturing Buyer: *Lisa McDowell*
Marketing Manager: *Holly Stark*

© 2004 Pearson Education, Inc.
Pearson Prentice Hall
Pearson Education, Inc.
Upper Saddle River, NJ 07458

Printed in the United States of America
10 9 8 7 6 5 4 3 2 1

ISBN 0-13-061977-9

Pearson Education Ltd., *London*
Pearson Education Australia Pty. Ltd. *Sydney*
Pearson Education Singapore, Pte. Ltd.
Pearson Education North Asia Ltd., *Hong Kong*
Pearson Education Canada, Inc., *Toronto*
Pearson Educación de Mexico, S.A. de C.V.
Pearson Education—Japan, *Tokyo*
Pearson Education Malaysia, Pte. Ltd.
Pearson Education, Inc., *Upper Saddle River, New Jersey*

Dedication

To my parents
and the educational assistance provided
by Hampden-Harvard tradesmen,
granite cutters from Barre Vermont,
and New England watermen.

Contents

Preface

Motivation and Intent

Bioengineering enrollments have recently soared. Indeed, 96 freshmen enrolled in the Spring 2003 course entitled "Introduction to Biomedical Engineering" at Carnegie Mellon. This course was the first required offering in a new double major at Carnegie Mellon, and intended to be deep enough to be on par with other first courses in traditional engineering majors.

Many excellent books exist on bioengineering topics. However, many require more advanced mathematical or biological expertise than freshmen or even sophomores possess. This book was written for freshman and sophomore engineers, and with curious colleagues in mind who desire to see the field's breadth and some depth. It is intended to provide a cross section of material.

Design of the Text

I tried to work backwards from the length of a semester and consider how students with a lot of demands on their time deal with information. Consequently, many good things were left out in order to produce 15 concise topics. The biomolecular, mechanical, materials, and electrical examples have proven to engage the students. One major surprise was that although the magnetic resonance material is quite challenging in that many layered concepts and phenomena have to be bundled to see how NMR/MRI works, the students responded best to this topic in all three semesters I have taught this course.

The book was also designed to be honest with students. If you want to really *do* bioengineering, it is a good idea to at least appreciate the "bio" part. Put another way, not all bioengineers will be gene cloners, but there is some aspect of cell biology or physiology that we all get to know well enough so that we may effectively communicate with our stakeholders and collaborators.

ABET and other specialists have advised that we elevate the social and political consciousness of our engineering graduates. Therefore, in some sections historical snippets are provided. The intent was not to waste space and placate those giving that advice. Rather, the history of bioengineering has much to be proud of and it is important to let our students know that their future work can also make a difference. Therefore, items such as Space Race Antibiotics (Chapter 0) and the long history of sugar (Chapter 7) appear in the text. The text also attempts to rationalize their future study in an effort to support the curriculum of your program and to build their overall view earlier in their education.

Lastly, bioengineers *are* engineers, and all engineers share some problem-solving concerns and motifs such as contending with trade-offs, optimization, and scaling. Because the course this text supports is supposed to be "engineering" vs. "a survey of what bioengineers do for a living," I attempted to illustrate common facets of engineering analysis through biological and medical examples. The level of analysis was placed somewhere between superficial and so hard for a freshman that the problem was forgotten in the midst of trying to deal with the math. Many homework problems are experiential in tone (pace optimization in hiking, drug candidate evaluation, bio space flight mission design, etc.) in an effort to make them interesting. I will leave it to you to determine why alcohol ingestion and metabolism was used as the pharmacokinetics example.

Textbook versus Web

In addition to the written text and solution manual, some web-based materials were developed for those who find the web a useful tool. First, a large collection of hyperlinks to sites that profile Nobel prize recipients and other potentially interesting sources of supplementary information have been assembled. Second, some basic animations and simulators are included that can be used in lectures or run by the students when they study. Third, slides that follow the book's chapters and a file with all the graphics used are available to the instructor. Currently, Prentice Hall will mount these materials on a web site and provide "keyed" access so the instructor can download them independently for customization efforts or instead, send students directly to the resources.

What Students Need to Know

My first polls on the prior experience of freshmen revealed that 60–80% claimed that they were exposed to enough biology beforehand to know what a gene and enzyme are. This year the number was 92%. Despite the positive upward trend, quite a few students still originate from high schools that quit after the tadpole-to-frog story. Thus, Chapter 1 is meant to be a refresher or equalizer. Beyond that, the text and problems encourage the students to apply the physics and math knowledge that they acquired or are concurrently learning. The majority of the problems (e.g., ligand binding analysis) are algebraic in nature. The calculus content is limited, and to provide for reinforcement, the particulars are "recycled" a number of times. Overall, a derivative is determined a few times, and the integration of dx, dx/x, and $dx/(a + bx)$ is performed within the context of "separation of variables" to solve some problems. Integration factors are introduced only once to challenge the more advanced students in the class. Most students indicated that they were pleased to apply some of the math they have labored to learn.

One Recipe for Running a Course

While many good things were left out, there still probably is more material than can be covered in one course, which hopefully will provide flexibility. Assuming an enrollment that is equally partitioned between students interested in chemical, mechanical, materials, and electrical facets, I have presented four units that address the range of potential interests. Also, I assume there are some things that all bioengineers should know about each other in addition to the basics of modern biology. I cover Chapters 0 and 1 in two lectures to equalize the biology. Part I covers Chapters 1–4 (Parts vs. Systems View of Biosystems). Chapters 5 and 6 followed by selected examples from Chapters 7 and 8 comprise Unit 2 (Molecular & Cellular Technology & Basic Informatics). Unit 3 (Mechanics & Materials) covers biomechanics (Chap 10) and portions of Chapters 11 and 12 to show the union of biomechanics and bio-materials by highlighting shear stress, clotting, and an overview of surface treatment strategies. Unit 4 exposes the students to signal processing and in the process, gives them a feel for how MRI works. Other instructors will have other circumstances so it is hoped that some chapters will fill in the backgrounds of diverse students or serve as a platform for providing additional depth based on the instructor's expert-ise and interests. The resources provided on the website may enable some deeper excursions on particular topics. For example, links are provided to workshop reports and three-dimensional displays of protein structures.

MICHAEL M. DOMACH
Carnegie Mellon University

Overview
of Bioengineering
and Modern Biology

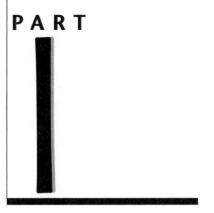

What is Bioengineering?

CHAPTER

0

0.1 | Purpose of This Chapter

Although you are a new student to this subject, you may have some well-informed, initial ideas on what bioengineering is and why it interests you. Indeed, breaking news on artificial heart transplants may have initially sparked your curiosity about this subject. Obviously with all those wires, moving parts, and complicated diagrams, engineers must have been involved somehow. The challenges and the rewards of improving human health may have also piqued your interest.

In this chapter, a general picture of engineering will be provided first. Then the types of work performed by those in the different bioengineering subfields will be described in an attempt to provide you with a complete picture. Throughout this chapter, links to subsequent chapters will be made in an attempt to provide a logical path for study.

0.2 | Engineering versus Science

Before considering what bioengineering is, it is useful to first consider what *engineering* is. A senior engineering professor once said, "The difference between engineering and science is all engineering equations have dollar signs in the denominator." While terse, there is a lot of truth to that statement, and such truths can be hard to extract from more lengthy and philosophical explanations. Engineers basically understand well the tools science and mathematics provide, and utilize them to solve problems for public and economic gain. Scientists, on the other hand, strive to create new knowledge about how things work and in the process, a language is developed to describe function and to use when debating ideas. Thus, the "dollar sign" reference, while an oversimplification, indicates the scientific knowledge has been reduced to practice and some resulting utility can be gauged, in part, by using a financial yardstick.

Note that despite the best intentions of the senior professor to simplify things, the difference between engineering and science does not imply that engineering is uncreative, solely financially driven, and only entails the use of established knowledge in handbooks. Scientists continually pose new questions about how things work to insure that understanding improves. The equally challenging question engineers ask is "Does a *new* solution to a current or foreseen problem exist that is economically sound, environmentally benevolent, and user friendly?" As Chapters 3 and 7 illustrate, engineers contend with a wide set of interacting issues when problem solving.

0.3 | Bioengineering

For most professionals, "bioengineering" is the catchall term for a variety of engineering specializations. Such catchalls are omnipresent in many disciplines. For example, a professor in an oceanography department may have a passion for long shore currents. Another professor in the same oceanography department may be fascinated by the natural history of a South Pacific mollusk. Thus, one can either be a physical oceanographer or an expert in ocean-dwelling invertebrates. While disparate, each specialization is germane to understanding how the seas function as a combined physical and living system. So while somewhat arbitrary, there are two major subfields in bioengineering: **biochemical** and **biomedical engineering**. Practitioners in either subfield have a solid to excellent understanding of modern biology and cell or human physiology. The problems they tackle and the base of scientific knowledge that guides their problem solving distinguishes them from mechanical, electrical, and other engineers. What, in turn, differentiates the two subfields is described next.

Biochemical engineering.   Biochemical engineers typically conceive and build processes that convert natural materials such as sugars into important molecules such as ethanol or therapeutic proteins like the blood clot-dissolving drug, streptokinase. The agent that transforms a raw material into a useful product is often a living cell. As you will see in Chapters 3, 5, and 8, microorganisms and other cells continually conduct a multitude of enzyme-catalyzed chemical reactions. Biochemical engineers thus strive to harness the synthetic capabilities of cells. In the process, genetic instructions within cells may be strategically modified such that more raw materials are diverted to the desired end-product. Based on this definition, one could perhaps claim that biochemical engineering was one of the earlier engineering disciplines because Mesopotamian artifacts from more than 6 thousand years ago display the seal of workers in front of a brewing vat.

Despite the possibility of tracing the roots of biochemical engineering back to Mesopotamia or earlier, contemporary biochemical engineering is viewed to have been founded shortly after another well-recognized engineering field, **chemical engineering**. Chemical engineering was officially recognized as an engineering field in the early 1900s. Contemporary biochemical engineering is closely affiliated with chemical engineering due to the discovery of the antibiotic, penicillin. **Penicillin**, which was discovered in 1929 by Alexander Fleming, is a molecule that is naturally produced by the *Penicillium notatum* mold during a particular stage in its growth.

Many cells and plants produce chemical compounds that aid in the defense of their environment against microbial or insect invaders. Fortunately, some of these molecules such as penicillin have been found to help combat human diseases.

During World War II, many scientists and chemical engineers working within American universities and pharmaceutical companies dramatically increased the scale and yield of penicillin-producing processes. Treating the infections incurred by growing numbers of battlefield casualties was a major motivation for their combined and accelerated effort. Large-scale processes were developed that used stirred tanks (fermenters) with controlled nutrient addition and oxygen supply to control extracellular conditions and growth. Penicillin production was a major technological stride because most prior chemical engineering processes did not use living materials to conduct the desired chemical transformations, and the tools biologists employed were adequate for only small-scale research efforts. Thus, ways had to be devised to provide for "life support" on a large scale while yet providing the right conditions and stimuli for convincing the living material to carry out the desired chemistry instead of other alternatives. These engineering efforts were performed in parallel to mutagenesis research programs that sought to develop strains of *Penicillium notatum* that produced even greater amounts of the antibiotic. Figure 0.1 displays some images of the early history of antibiotic production in the United States.

The success with generating large quantities of the "miracle drug," penicillin, can thus be viewed as the genesis of the modern biochemical engineer. These new engineers learned biology along the way and they soon afterwards develop biological-based processes for producing other drugs (e.g., steroids) and antibiotics. Thus, chemical engineering process knowledge was now fruitfully fused with life science and chemistry.

As with all engineering efforts, interesting links exist with historical events, which illustrates that science and engineering do not occur in a cultural or political vacuum. Indeed, treating large numbers of battlefield casualties provided a major impetus to develop and produce antibiotics on a large scale. However, because antibiotics were so new in the 1940s and 1950s, and their effects on combating infection were so startlingly successful, the science and engineering prowess associated with their discovery and production became mingled with national pride and prestige.

To illustrate, you may have heard of the "Space Race." In the 1960s and 1970s, the United States was competing with the former Soviet Union for technical and economic superiority. Being the first to land a human on the moon was regarded as a major technological and moral victory. However, other engineering endeavors were also viewed with the same lens during the Cold War. To illustrate, you may have heard of the "Space Race." Throughout the 1950s and 1970s the United States competed with the former Soviet Union for technical and economic superiority. Being the first to land a human on the Moon was regarded as a major technical and moral victory. However, biochemical and other engineering endeavors were also viewed through the same competitive lens. As a trip to the library will reveal, articles published in professional journals tended to downplay the originality and potency of "Iron Curtain" antibiotics (e.g., Industrial and Engineering Chemistry, vol. 51, Sept. 1959, p. 1087).

Today, biochemical engineers continue to work on designing and operating production processes. Some processes produce small quantities of potent, valuable

FIGURE 0.1   Early illustrations of antibiotic-production processes. (Top, left) The gentleman on the left in the foreground is Selman Waksman, the Ukraine-born discoverer of streptomycin. He received the Nobel prize in medicine in 1952. Waksman is touring a Merck facility as a workman passes overhead (circa 1946 and courtesy of John Higgins, Merck). (Top, right) The early fermentation tank for producing penicillin at Merck engulfs a workman (courtesy of Merck Archives). (Bottom). A 1940s era facility devoted to antibiotic production.

products. Other processes can be quite large in physical scale as the world's largest fermenter attests in Figure 0.2. Modern biochemical engineers also view the biochemical and genetic networks inside cells as raw blueprints, which can be modified and engineered such that cells will produce more of a desired molecule. Thus, many biochemical engineers are well versed in molecular biology and sometimes it is hard to distinguish them from molecular biologists and biochemists when they discuss their work. Examples of cell-scale work will be provided in Chapter 8 in the section on **Metabolic Engineering**. Designing and operating the equipment that separates the desired molecule from the myriad of others is also under the jurisdiction of the biochemical engineer.

Biomedical engineering.   Simply put, biomedical engineers design devices and methodologies that enable the detection, diagnosis, management, and/or elimination of disease. Health care deliverers and patients are the ultimate users of the efforts of biomedical engineers, although engineers can often be found in the health care delivery system where they, for example, assist surgeons with the implantation of devices that correct irregular heart rhythms.

   This definition covers a lot of ground, so some examples may help to visualize the different incarnations. Biomedical engineers who possess a strong foundation in mechanical engineering can develop limb prostheses or rehabilitation systems.

FIGURE 0.2   An example of the large-scale nature of some industrial fermentation processes. When erected in 1978, the "bioreactor" towered 200 feet high. This photo appeared in *Chemical and Engineering News* (September 18, 1978), and is used with permission granted by the Corporate Communications branch of Imperial Chemical Industries.

Others can improve human-work environment/machine interfaces to increase safety and to decrease stress. One example is provided in Chapter 10, which introduces biomechanical engineering. Those who understand fluid flow mechanics and human and cell physiology can work with clinicians to design patient life support systems such as the heart-lung bypass, blood pumping, and oxygenation system shown in Figure 0.3. An introduction to blood flow and medical device design is provided in Chapter 11. Other practitioners that understand material synthesis and properties as well as the body's immune and wound healing systems design and produce materials that can be implanted into the body to remedy problems such as joint degeneration, broken bones, etc. Some basic concepts and examples that illustrate this specialization, which is called biomaterials engineering, are covered in Chapter 12. Those that have knowledge of electromagnetism and the nature of some diseases that are more treatable with early diagnosis contribute to the design of noninvasive diagnostic technologies such as magnetic resonance imaging (MRI). The fundamentals and engineering involved in MRI will be covered in Chapter 14.

Although these examples are varied and many more exist, one common denominator should be apparent. All modern biomedical engineers possess a solid foundation in life science knowledge and substantial depth in some particular aspects. Another facet, which may not be obvious from the broad definition, is biomedical engineers have rather unique stakeholders and responsibilities compared to other engineers.

The product of biomedical engineering must be beneficial to both a patient and health care deliverer. This means that in addition to the product functioning well, a nurse, physician, surgeon, or technologist must be able to use and trust the MRI, patient monitor, or other engineered product when it is used repeatedly and

FIGURE 0.3 Schematic of what a Heart Lung Machine (HLM) accomplishes. The pump takes over the job of the heart, and first pumps the blood through the oxygenator. The oxygenated blood is then returned to the patient to support their tissues and organs. The HLM has five precision blood pumps. The machine also has a heater/cooler system and controllers that regulate patient temperature and gas supply.

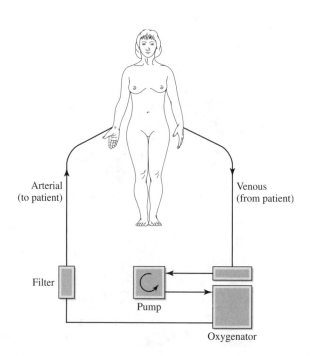

sometimes under stressful circumstances. If the product is difficult to use, it will not be used and its benefits never realized. Worse yet, failures or misuse could lead to unintended violations of one of the basic tenets of medicine as set forth by the famous Greek philosopher, Hippocrates: *"As to diseases, make a habit of two things—to help, or at least **do no harm**"* (Hippocrates, *The Epidemics)*. The full text of the **Hippocratic Oath** is included at the end of this chapter. Some aspects of the trade-offs in medical product development are discussed in Chapter 7 within the context of a blood glucose-measuring device.

Another important stakeholder is the **United States Food and Drug Administration (USFDA)**. Although many engineers embrace certain federal and industry safety standards, such as building bumpers that can withstand 5-mile-per-hour impacts, gaining USFDA approval requires that extensive testing and application processes be successfully surmounted. The activities of the USFDA have resulted in a largely safe and rapidly responding medical system. Indeed, when a medical product gains USFDA approval, it is regarded as receiving one of the best endorsements available in the world. However, the regulatory process introduces unique challenges to product design and development. Designs and methods have to be "nearly perfect" the first time because once the commitment is made to the time-consuming testing and approval processes, the design is "frozen." Any revisions will have to go through approval processes; hence, unanticipated problems are costly and time-consuming matters. Thus, a biomedical engineer's creativity must be applied to medical and engineering problem solving *and* anticipating usage problems or undesirable side effects that arise from biological phenomena.

Biomedical engineering emerged as a discipline a little later and in a different manner compared to its "cousin", biochemical engineering. Dispersed attempts at formally organizing a biomedical engineering discipline within the United States began in the late 1960s. The later genesis of biomedical engineering may be attributed to the lack of a "crystallizing nucleus" such as the "penicillin campaign" that initiated the birth of biochemical engineering. Rather, biomedical engineering coalesced gradually at first, and then the pace accelerated. The same might be said for chemical engineering, which began as "recipe-driven, industrial chemistry." Thereafter, new chemistry knowledge emerged along with an increased demand for innovation from the petrochemical industry. These factors spurred the development of chemical engineering as a discipline, which entailed integrating engineering analysis and design skills with chemistry knowledge.

In the 1960s and 1970s, biomedical engineering was populated and organized primarily by electrical engineers and physicists. These specialists were engaged because radiation- and electrical device-based advancements were increasingly used as patient treatment and monitoring technologies. As the medical world began to embrace the prospect of an artificial heart and the utility of other implants, members of other established engineering disciplines increasingly devoted their efforts to problem solving in the medical domain. Mechanical and material engineers, for example, contributed to the improvement of artificial organs. Concurrent advances in pharmaceuticals then motivated chemical engineers to apply their knowledge to developing mathematical models that predict how drugs and substances accumulate and distribute in the human body.

Over the last decade, spectacular advances in genetics and cell biology have been occurred, which have had an impact on engineering practice and organization.

One impact is more engineers have been drawn into the world of cells and tissues in order to explore the practical applications of the advances. These bioengineers call themselves either **metabolic engineers** or **tissue engineers**, depending on whether they seek to produce an economically important molecule or solve a tissue-related, medical problem. The fields of **metabolic engineering** and **tissue engineering** are covered in Chapter 8, and an example of a result from tissue engineering work is shown in Figure 0.4. Another impact is the increased adoption of biomolecular and cellular science into the scope of bioengineering has blurred somewhat the distinction between some biochemical and biomedical engineers. A significant subset of biochemical and biomedical engineers now routinely grow cells and manage cellular control "circuits" in order to produce a useful molecule or a functional ensemble of cells (i.e., tissue replacement). Yet another impact is a stronger scientific base now exists for pursuing other biomedical engineering endeavors such as the design of drugs and hybrid artificial organs. From the perspective of a college student, perhaps the most immediately relevant impact is many academic institutions and professional engineers have recognized the high potential of the advance's applications, and the response has been to offer formal academic programs in bioengineering and biomedical engineering.

Because of the distinctive presence of life science in bioengineering and its increasing emphasis, this book begins with a review of cellular and biochemical science in Chapter 1. Thereafter, the operation of control mechanisms and the quantitative descriptions are covered in subsequent chapters. Material and energy requirements to support a living cellular or human-scale system are presented in Chapters 2 and 4.

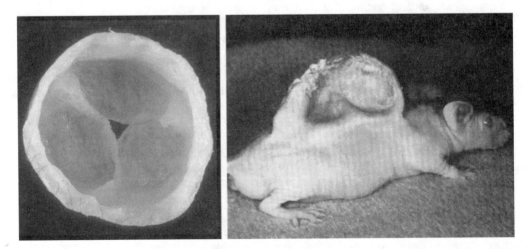

FIGURE 0.4   Examples of recent tissue engineering efforts. On the left, cells were coaxed to grow into a prototype for a natural human heart valve replacement. The human ear grown on the mouse on the right was created using an implanted scaffold to support and steer the proliferation of human cartilage cells. Because the specially bred mouse lacks a fully functional immune system, the human cells are supported by the mouse's circulatory system as opposed to being rejected and destroyed.

0.4 | Career Opportunities

Even though graduation may appear to be quite distant in your future, some possible career paths can be sketched that may help you plan your education. With a B.S. degree, you could be situated to join a pharmaceutical company as a process engineer. That means you may be responsible for designing and running the equipment that produces an antibiotic. Alternately, you could become involved with producing a quantity of a new drug so that clinical trials can begin. Another path is to join a clinical engineering group in a hospital. There, your efforts would be aimed at insuring that the patient monitoring and diagnosis equipment functions well, and your expertise would help illuminate where improvements are desirable. There are many possibilities, but these illustrate how you can focus your study on molecular science and process technology or bio-instrumentation to achieve different outcomes.

Many B.S. graduates with a major or minor in bioengineering elect to continue their education. The hallmark of the typical bioengineering curriculum is the presence of significant science content, which tends to attract curious students. Thus, it is not surprising that many bioengineering students attend graduate school after completing their B.S. degree. Students who obtain M.S. or Ph.D. degrees in engineering after 2 or 4.5 years of postgraduate study, respectively, become equipped to perform the basic research that leads to advancements in drug therapies, tissue engineering, rehabilitation engineering, and other applications. Jobs exist within large companies, small new companies, and large medical research establishments such as the University of Pittsburgh Medical Center. While more schooling may not seem like an attractive option now, bear in mind that most well-regarded, engineering graduate programs will at least provide a tuition scholarship for postgraduate education and research. Many graduate programs will also provide a stipend that covers your living expenses.

The clinical researcher and physician ranks have recently become increasingly populated with people who earned B.S. degrees in engineering. This reflects how the interests of engineering students have changed and their increased enrollment into medical schools or M.D.-Ph.D. programs after graduation. **M.D.-Ph.D. programs** are often financially supported by grants from the National Institutes of Health. These grants are competed for and awarded to universities with meritorious educational and research programs. The financial resources are used to provide tuition scholarships and stipends for students that combine traditional medical school with Ph.D. level research in life science or engineering. Typically after two years of medical school, students devote most of their time to study and research in an allied graduate science or engineering program, while remaining attached to the medical school by, for example, participating in half day long clinical clerkships each week. After obtaining the Ph.D. (three to five years), students complete their medical training. Afterwards, a research or a medical residency position is sought. The benefits for those who take the M.D.-Ph.D. path in bioengineering is that advanced engineering or science credentials are achieved, as well as the ability to work directly with patients in a clinical setting. Often, M.D.-Ph.Ds work with patients in a clinic like a regular physician, *and* they also manage a research group. The financial support provided by many M.D.-Ph.D. programs can also ease the loan burden on a student. The time commitment, however, is significant.

0.5 | Introductory Note

HIPPOCRATES, the celebrated Greek physician, was a contemporary of the historian Herodotus. He was born on the island of Cos between 470 and 460 B.C., and belonged to the family that claimed descent from the mythical AEsculapius, son of Apollo. There was already a long medical tradition in Greece before his day, and this he is supposed to have inherited chiefly through his predecessor Herodicus; and he enlarged his education by extensive travel. He is said, though the evidence is unsatisfactory, to have taken part in the efforts to check the great plague which devastated Athens at the beginning of the Peloponnesian war. He died at Larissa between 380 and 360 B.C.

The works attributed to Hippocrates are the earliest extant Greek medical writings, but very many of them are certainly not his. Some five or six, however, are generally granted to be genuine, and among these is the famous "Oath." This interesting document shows that in his time physicians were already organized into a corporation or guild, with regulations for the training of disciples, and with an esprit de corps and a professional ideal which, with slight exceptions, can hardly yet be regarded as out of date.

One saying occurring in the words of Hippocrates has achieved universal currency, though few who quote it today are aware that it originally referred to the art of the physician. It is the first of his "Aphorisms": "Life is short, and the Art long; the occasion fleeting; experience fallacious, and judgment difficult. The physician must not only be prepared to do what is right himself, but also to make the patient, the attendants, and externals cooperate."

0.6 | The Oath of Hippocrates

I SWEAR by Apollo the physician and AEsculapius, and Health, and All-heal, and all the gods and goddesses, that, according to my ability and judgment, I will keep this Oath and this stipulation—to reckon him who taught me this Art equally dear to me as my parents, to share my substance with him, and relieve his necessities if required; to look upon his offspring in the same footing as my own brothers, and to teach them this art, if they shall wish to learn it, without fee or stipulation; and that by precept, lecture, and every other mode of instruction, I will impart a knowledge of the Art to my own sons, and those of my teachers, and to disciples bound by a stipulation and oath according to the law of medicine, but to none others. I will follow that system of regimen which, according to my ability and judgment, I consider for the benefit of my patients, and abstain from whatever is deleterious and mischievous.

TABLE 0.1

Some Prominent Engineering and Science Societies

American Association for the Advancement of Science	American Association of Physicists in Medicine
American College of Clinical Engineering	American Institute for Medical and Biological Engineering
American Institute of Chemical Engineers	American Institute of Ultrasound in Medicine
American Medical Informatics Association	American Society for Artificial Internal Organs
American Society for Engineering Education	American Society for Healthcare Engineering
American Society of Agricultural Engineers	American Society of Biomechanics
American Society of Echocardiography	American Society of Mechanical Engineers
Association for the Advancement of Medical Instrumentation	BioMedical Engineering Society of India
Biomedical Engineering Society	Biophysical Society
Canadian Society for Biomechanics	Computers in Cardiology
Controlled Release Society	Danish Society for Biomedical Engineering
European Society for Engineering in Medicine	European Society of Biomechanics
Institute of Biological Engineering	Institute of Electrical and Electronics Engineers • IEEE Engineering in Medicine and Biology Society
Institute of Physics and Engineering in Medicine	International Biometric Society
International Council of Science	International Federation for Medical and Biological Engineering
International Lung Sounds Association	International Organization for Medical Physics
International Society for Computational Biology	International Society for Magnetic Resonance in Medicine
International Society for Optical Engineering	International Society of Biomechanics
International Society of Electrophysiology and Kinesiology	Radiological Society of North America
Rehabilitation Engineering and Assistive Technology Society of North America	Society for Biomaterials
Society for Computer Applications in Radiology	Society for In Vitro Biology
Society for Mathematical Biology	Society for Physical Regulation in Biology and Medicine
Society of Nuclear Medicine	The Institution of Engineers, Australia • College of Biomedical Engineering
Tissue Engineering Society International	

I will give no deadly medicine to any one if asked, nor suggest any such counsel; and in like manner I will not give to a woman a pessary to produce abortion. With purity and with holiness I will pass my life and practice my Art. I will not cut persons laboring under the stone, but will leave this to be done by men who are practitioners of this work. Into whatever houses I enter, I will go into them for the benefit of the sick, and will abstain from every voluntary act of mischief and corruption; and, further, from the seduction of females or males, of freemen and slaves. Whatever, in connection with my professional service, or not in connection with it, I see or hear, in the life of men, which ought not to be spoken of abroad, I will not divulge, as reckoning that all such should be kept secret. While I continue to keep this Oath unviolated, may it be granted to me to enjoy life and the practice of the art, respected by all men, in all times. But should I trespass and violate this Oath, may the reverse be my lot.

0.7 | The Law of Hippocrates

1. Medicine is of all the arts the most noble; but, owing to the ignorance of those who practice it, and of those who, inconsiderately, form a judgment of them, it is at present far behind all the other arts. Their mistake appears to me to arise principally from this, that in the cities there is no punishment connected with the practice of medicine (and with it alone) except disgrace, and that does not hurt those who are familiar with it. Such persons are the figures which are introduced in tragedies, for as they have the shape, and dress, and personal appearance of an actor, but are not actors, so also physicians are many in title but very few in reality.

2. Whoever is to acquire a competent knowledge of medicine, ought to be possessed of the following advantages: a natural disposition; instruction; a favorable position for the study; early tuition; love of labor; leisure. First of all, a natural talent is required; for, when Nature leads the way to what is most excellent, instruction in the art takes place, which the student must try to appropriate to himself by reflection, becoming an early pupil in a place well adapted for instruction. He must also bring to the task a love of labor and perseverance, so that the instruction taking root may bring forth proper and abundant fruits.

3. Instruction in medicine is like the culture of the productions of the earth. For our natural disposition, is, as it were, the soil; the tenets of our teacher are, as it were, the seed; instruction in youth is like the planting of the seed in the ground at the proper season; the place where the instruction is communicated is like the food imparted to vegetables by the atmosphere; diligent study is like the cultivation of the fields; and it is time which imparts strength to all things and brings them to maturity.

4. Having brought all these requisites to the study of medicine, and having acquired a true knowledge of it, we shall thus, in traveling through the cities, be esteemed physicians not only in name but in reality. But inexperience is a bad

treasure, and a bad fund to those who possess it, whether in opinion or reality, being devoid of self-reliance and contentedness, and the nurse both of timidity and audacity. For timidity betrays a want of powers, and audacity a lack of skill. They are, indeed, two things, knowledge and opinion, of which the one makes its possessor really to know, the other to be ignorant.

5. Those things which are sacred, are to be imparted only to sacred persons; and it is not lawful to impart them to the profane until they have been initiated into the mysteries of the science.

REFERENCES

Functional living trileaflet heart valves grown in vitro. Hoerstrup, S.P., Sodian, R., Daebritz, S., Wang, J., Bacha, E.A., Martin, D.P., Moran, A.M., Guleserian, K.J., Sperling, J.S., Kaushal, S., Vacanti, J.P., Schoen, F.J., Mayer, J.E. *Circulation*. 2000. 102 (suppl. III): 44-49.

Transplantation of chondrocytes utilizing a polymer-cell construct to produce tissue-engineered cartilage in the shape of a human ear. Cao Y., Vacanti J.P., Paige K.T., Upton J., and Vacanti C.A. *Plast Reconstr Surg*. 1997. 100(2): 297-302.

1

Cellular, Elemental, and Molecular Building Blocks of Living Systems

1.1 | Purpose of This Chapter

About the time the American War between the States was being fought, **Louis Pasteur, Robert Koch** (http://www.nobel.se/medicine/laureates/1905/koch-bio.html), **Jean Jacques Theophile Schloesing** (http://dev.asmusa.org/mbrsrc/archive/SIGNIFICANT.htm), and others demonstrated that small, unicellular organisms exist. Moreover, their work showed that these cells are agents of infection, alcohol production in wine, and global nitrogen cycling.

Roughly a century later, western Pennsylvania native **Herbert Boyer** (see http://www.accessexcellence.com/AB/BC/Herbert_Boyer.html), and his colleague, **Stanley Cohen**, inserted the genetic material from a toad into a common intestinal bacterium. The result was that the bacteria's synthetic machinery produced a molecule found in a vastly different species. The intriguing prospect that bacteria could be used as factories for producing complex biological molecules with therapeutic value, such as insulin, led Boyer and Robert Swanson to form the company **Genentech**. Much has happened in a short time and information on molecular biology continues to expand at a bewildering pace.

Whether a bioengineer is involved with designing an artificial heart or developing a production process for cell-derived insulin, some knowledge of modern molecular and cell biology is needed to interact with other experts in order to render the best possible design. Moreover, such knowledge is needed to fully appreciate the ultimate positive (and potential negative) impacts of the product or system on a patient's health. Indeed, ample examples exist that illustrate that adverse consequences can result when biological factors such as immune system reaction are not considered or rapidly responded to when new scientific information becomes available. The rapid breakdown of a jaw joint implant and resulting immune system activation provides one notable historical example (http://www.fda.gov/bbs/topics/NEWS/NEW00249.html).

In an effort to review your prior studies and to provide a basis for understanding the topics in this text, this chapter will provide an overview of the basic vocabulary and philosophies of biological science. The focus will be at the cellular level because cells are the building blocks of living systems. Additional details will appear in subsequent chapters to allow for applications. In general, mastering the vocabulary and concepts of biological science will empower you to interpret and utilize scientific findings in engineering efforts. First, origins and divergence are covered. Thereafter, the elemental and molecular components of cells will be described. This chapter concludes with a summary of physiology, viruses, and prions in order to provide a complete view of the diversity within the world of replicating, biological systems. Follow-up study of biochemistry and cell physiology will empower you further.

1.2 │ Origins and Divergence of Basic Cell Types

Biological science has traditionally provided useful ways to classify living systems. These classifications provide a way to group similar cells and organisms. Through classification, a vocabulary is also provided so different people can discuss issues in a coherent manner. This is akin to naming clouds as nimbus or cumulus. Such terms, while arbitrary, allow one to succinctly describe and discuss local weather with someone else at another location, given that the terms and meanings are agreed upon. The science of classification is called **taxonomy**; the word is derived from the Greek root *taxo* (to organize or put into order). The organization of life by ancestor-descendent (evolutionary) relationships is called **phylogeny**.

More recently, biology has entered an era of molecular determinism. It is now possible to identify what molecules are engaged in transmitting parental traits to off-spring, guiding the development of an embryo to a complex organism, etc. Additionally, elucidating the different ways by which molecules function *together* to elaborate reproduction, repair injuries, provide locomotion, and confer other salient features of living organisms is of keen interest. The established molecular basis, which describes how cells function at the mechanistic rather than the behavioral level, has been used to challenge and alter prior classification schemes. We will first review the current classifications of cells. After discussing elemental and molecular constituents of cells and information storage and utilization an example of how the classification scheme was developed will be provided.

The current classification of cell types and how they are thought to be evolved from a common ancestor is shown in Figure 1.1. The three types are **Bacteria, Archaea,** and **Eucarya**. Bacteria are unicellular organisms that are capable of reproducing. They range in size and shape; a typical dimension is 1 micron (10^{-6} meter). Bacteria are found in a wide range of environments and exhibit different capabilities. The rod-shaped bacterium, *Escherchia coli*, for example, can use just glucose and inorganic salts as raw materials for growth and multiplication. Figure 1.2 shows what these bacteria look like. Other bacteria exist in parasitic or symbiotic relationships. They must acquire some raw materials from the host or symbiont for growth and multiplication.

Bacteria can be agents of disease. For example, a streptococcus infection results in "strep throat" and a week off from school or work. Worse yet, George Washington is believed to have perished from a streptococcus infection because the antibiotics we now take for granted were not available in the 1700s. However, without the activity of bacteria, life as we know it would be quite different. Bacteria break down pollutants and organic matter in the environment, thereby allowing elements

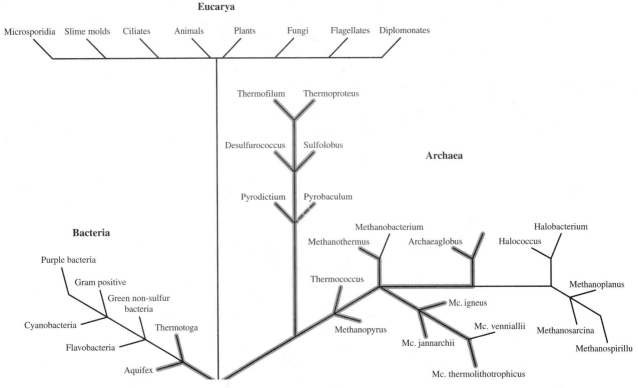

FIGURE 1.1   Cellular life is viewed as splitting into three groups and originating from a common ancestor.

and chemical building blocks to reenter food chains and nutrient cycles. Additionally, some bacteria in the digestive systems of animals convert ingested food molecules into essential vitamins. Indeed, bacteria are a large fraction of the functional biomass found on the living surface of the Earth. Finally, bacteria are among the workhorses of the biotechnology industry. Through genetic manipulation, they can be programmed to produce pharmaceutical products such as insulin. **Biotechnologists** and **biochemical engineers** work on harnessing this productive potential.

The **Archaea** are similar to bacteria in many respects. They are about the same size and grow on a number of raw materials. One difference is they tend to be found

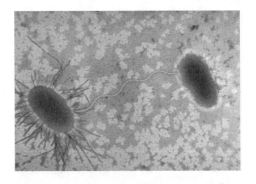

FIGURE 1.2 A transmission electron micrograph of the bacterium, *Escherichia coli (E. coli).*

in more extreme environments. These environments include hot springs and acidic waters. These environments are thought to resemble those found in the early days of the Earth, so Archaea are viewed as ancient remnants of the past. Their ability to prosper in extreme environments has also piqued interest about the possibility of life beyond the Earth. Because Archaea live in extreme environments, others view their proteins and other biological molecules as being potentially unique sources of catalysts and medicinal compounds. As will be detailed later in this chapter, the chemical composition of some molecules found in Archaea also differs from bacteria, which allows for their identification and definition in precise molecular biological terms.

Eucarya include the cells that compose our body. One distinguishing feature is these cells have compartments within them. Consequently, Eucarya are viewed as more complex and evolved than bacteria and Archaea. These compartments are called **organelles**. One important organelle is the **nucleus**, which houses the DNA molecule. The DNA molecule(s) encodes all the information for building the cell and conferring it with function.

In animals, cells can either be **primitive** or **terminally differentiated,** or fall in between these two extremes. The former means that when certain chemical or other signals are received, a particular cell can be activated to form another type of cell. **Bone marrow stem cells** provide one example. When quiescent, they exist in modest number in the bone marrow mass. When certain signals are transmitted and received, however, these stem cells start replicating and transforming themselves into the cells of the immune system, red blood cells, or other cells that the body requires. The transformation into a specialty is called **differentiation**.

A terminally differentiated cell is one that has undergone the change in function and in the process, has lost the ability to replicate. The platelets found in the blood stream provide one example. Platelets participate in wound healing, which will be presented in more detail in Chapter 12, when biomaterial engineering is introduced.

The ability of some cells to differentiate has excited those interested in repairing or regenerating diseased organs. Many researchers envision that primitive cells can be activated and coaxed to divide and grow into a functional tissue or organ. This possibility is the basis for one line of research in the field, **tissue engineering**. Tissue engineering is the practice of growing cells in a controlled manner such that the spatial organization and metabolic attributes of healthy organs or tissues are attained. The example of using (presumed) stem cells to grow a replacement retina for vision-impaired patients will be provided in Chapter 8.

1.3   Elemental and Molecular Composition of a Cell

Elemental Abundance.   Cells and thus humans are mostly water. A typical cell is about 70 percent water and on a water-free basis, the mass is 50 percent by weight carbon. Nitrogen and phosphorous are the next most abundant elements; they typically constitute 14 and 4 percent of the water-free mass, respectively. That means for 1 g of cells, there is 0.3 g of water-free material and of that mass, carbon, nitrogen, and phosphorous are present as 0.15 g, 0.04 g, and 0.01 g, respectively.

Protein and Amino Acids Constituents. As is the case for the elemental composition, some molecules are more prevalent than others. Working again on a water-free basis, **protein** is the most abundant molecule in terms of mass percentage, and a typical value is 60 percent. A protein is a polymer of **amino acids** as shown in Figure 1.3. The polymer nature of the protein molecule means that individual acids are connected to make a long chain. The chain then folds upon itself to form an organized globular or rod-like structure. As will be discussed further in Chapters 3 and 5, the particular three-dimensional shape attained contributes immensely to the functionality of the protein.

As the structures in Figure 1.3 indicate, an amino acid derives its name from the fact that an **amino group** (NH_2) is attached to a carbon that is adjacent to an acidic (H^+ yielding) **carboxyl** moiety (COOH). What is linked to the NH_2-bearing carbon can vary, so a general amino acid is often denoted as R-CH(NH_2)-COOH, where the "R" symbol indicates that different combinations of elements can be attached to the defining portion of the molecule. There are twenty common amino acids, and so there are twenty different R groups. **Glutamate** is a very prevalent amino acid found in cells; it is also the active component of the seasoning, monosodium glutamate, which is often referred to as MSG.

Some cells can make all twenty amino acids. In other cases, some amino acids must be provided by food intake; hence, those amino acids are referred to as "**essential (or required) amino acids.**" As Table 1.1 summarizes, humans require nine amino acids to be provided by dietary intake. Rats and humans have the same amino acid requirements, whereas some bacteria can synthesize all twenty from sugars and inorganic salts, which raises an interesting question about the definition of an "advanced" versus a "simple" organism.

Within a cell or our bodies, proteins perform multiple functions. They can, for example, serve as structural elements, as represented by the collagen fibers in the

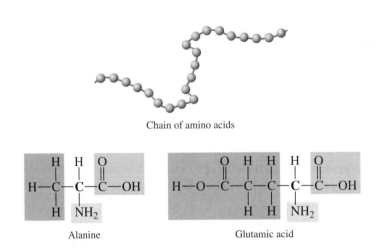

Chain of amino acids

Alanine

Glutamic acid

FIGURE 1.3 A protein chain consists of different connected combinations of twenty amino acids. All amino acids have an alpha carbon (**bold**), to which an acidic carboxyl group (COOH) and an amino group (NH_2) are attached. The other group attached to the alpha carbon varies and distinguishes one amino acid from another. Two common amino acids are alanine and glutamic acid.

TABLE 1.1

Essential vs. Nonessential Amino Acids	
Nonessential	Essential
Alanine	Histidine
Arginine	Isoleucine
Asparagine	Leucine
Aspartate	Lysine
Cysteine	Methionine
Glutamate	Phenylalanine
Glutamine	Threonine
Glycine	Tryptophan
Proline	Valine
Serine	–
Tyrosine	–

skin. As discussed in later chapters, proteins can also bind to other molecules to enable the turning on and off of different events. The acceleration (i.e., catalysis) of biochemical reactions is also mediated by special proteins called **enzymes**, which will be discussed further in later chapters.

1.4 | Molecules That Contain Information

The **deoxyribonucleic acid (DNA)** molecule is less abundant than protein (a few mass percent on a water-free basis), but its structure-endowed function is extremely important. The DNA molecule contains all the instructions for producing every molecule currently in the cell as well as those that could have some future use if environmental conditions change or differentiation occurs. Four bases are linked in varying patterns to build this biopolymer. The four bases are **adenine (A)**, **cytosine (C)**, **guanine (G)**, and **thymine (T)**. The DNA found in cells contains two strands, sometimes called a **duplex** or **double-stranded DNA**. Attractive forces between A and T, as well as G and C, hold the two strands together and contribute to establishing the α-helical, coil spring-like structure of DNA. The association of A (or G) on one strand with T (or C) on the other strand is called **base pairing**.

The sequence in which the A, C, G, and T bases appear on a strand is very important because of the existence of the **genetic code**. To illustrate how the genetic code works in conjunction with the composition of DNA, consider a protein composed of twenty amino acids. The amino acids must be linked in a particular order for the protein to fold properly and perform its function. This is accomplished by devoting a stretch of the DNA molecule for recording the recipe for this protein. This stretch containing the recipe is known as a (structural) **gene**. Within the stretch, the A, C, G, and T bases are linked according to a code to provide the instructions for linking the protein's amino acids in the correct order. The linkage is such that the information is read in a set direction much the way a cassette tape is "read" in a fixed direction so the recording and playback of music are consistent and intelligible.

How Information and Language Are Imbedded in DNA. Every three bases code for one amino acid; a set of three bases is called a **triplet**. Each amino acid has its own triplet(s), just like we have unique phone numbers. For example, GAA and GGG encode for the amino acids, glutamate and glycine, respectively. So if the first two amino acids that appear in the protein are glutamate and glycine, the bases will appear in the gene in the order: GAAGGG. Since there are twenty amino acids in the protein, there will be twenty triplets and thus a total of sixty bases are needed to code for the protein.

Basic information theory suggested why triplets are used and provided prior researchers with some hints as to what to look for when trying to crack the genetic code. When one has n letters to work with, the number of unique words consisting of a specified number of letters, m, is n^m. That is, when developing a language with a given alphabet, there is a maximum number of unique words of fixed length that can be assembled from the alphabet. To illustrate, consider an alphabet consisting of three letters, A, B, and C. The formula indicates that the number of two-letter words is $3^2 = 9$. We can work out the unique two letters words by hand to verify the formula; the words are

$$
\begin{array}{ccc}
AB & AC & BC \\
BA & CA & CB \\
AA & BB & CC \\
\end{array}
$$

When there are four letters (A, G, C, T) in the alphabet, there must be at least 20 words in order to encode for the 20 common amino acids. If two letter words were used (e.g., AA, AT, GC), we come up short because there are only $4^2 = 16$ unique words. If instead, three letter words were used to encode for the 20 amino acids, the language now has $4^3 = 64$ words, which is more than sufficient. The 44 extra words means that synonyms exist in the genetic language in that some amino acids may be encoded by more than one triplet. However, no two amino acids share a triplet. Some extra words can also be used for indicating where genes begin and end.

Mutation, Phenotype, and Genotype. Sometimes the code is altered when a cell makes a copy of its DNA during reproduction. The large number of bases in a DNA molecule and multitude of cell divisions that can occur over an animal's lifetime means that even if the error rate in the copying process is low, mistakes can occur. This is akin to airplane safety. Strict and effective maintenance standards exist, and the average person's odds of being in an accident are low, but the large number of passenger-miles still results in an accident now and then. Chemicals found in the environment or other factors may also modify the DNA molecule and thus the coding it contains. An alteration in an organism's genetic code is called a **mutation**.

When the code is altered, there may be no effect because a triplet may be changed into another that can code for the same amino acid. Thus, the same protein can be produced. For more radical changes in base sequence, the instructions may no longer faithfully produce a fully functional protein. However, even such a mutation may not necessarily result in the death of a cell or an animal. Rather, the ability to contend with only one situation may be impaired, while other functions can be fulfilled.

Because an offspring receives some DNA coding from the parent(s), a mutation can be inherited. Hemophilia, which, arises when the gene that codes for a blood clotting protein (factor VIII) is mutated, is a classic example. A person carrying that mutation can function fine until injured by a cut or bruised. Then, excessive bleeding occurs which can be lethal.

What traits an organism presents is called the **phenotype** and the traits are linked to the instructions encoded in the DNA. The raw instructions are, in turn, are called the **genotype**. Modern molecular biology tools have allowed researchers to unravel the genotype of organisms ranging from bacteria to humans. The existence of synonyms in the genetic code and other issues, however, has posed the challenging problem of relating an organism's phenotype to genotype. Research is now underway that is aimed at trying to more fully understand how genotype and phenotype are linked.

Using the Information in DNA.   The information present in the DNA molecule is "read out" via other molecular machinery. First, a photocopy of a gene is made, much like what you do when you go to the library. The full volume remains on the shelf and serves as an archival copy. You copy only the particular section of text you immediately need. The biomolecular "photocopy" is called **messenger RNA (mRNA)** and the copying process is referred to as **transcription**. More details on this will be provided in Chapter 5 when binding is covered in more mechanistic and mathematical detail.

The sequence information now copied on the mRNA molecule is then **translated** and utilized by large molecular complexes called **ribosomes**. The ribosomes first bind to the mRNA copy. Then the ribosomes move down the length of the mRNA molecule. At each triplet, a ribosome in a sense pauses and allows for the triplet-prescribed amino acid to be added to the growing protein molecule at precisely the right position. This process resembles how you read a sentence. Your eye moves from left to right and each word has particular significance to you. At the end of the sentence, all the words combine to convey a particular thought. Thus, the triplets are akin to words, and the sequence of amino acids that leads to a protein that executes a particular function is analogous to a complete sentence. The unit of sequence information that encodes for an amino acid is also called a **codon**.

Copying DNA in the Laboratory.   Through the study of cells, the enzymes that enable a cell to copy its DNA for passage on to a progeny cell have been well established. When a process such as DNA replication occurs within an *intact* cell, it is called *in vivo* DNA replication. Many aspects of how cells replicate their DNA can now be reproduced in the laboratory without intact cells. When a subset of cellular components is used in, for example, a test tube, to perform a task, the reproduced natural process is said to be performed *in vitro* (e.g., "*in vitro* DNA replication").

Three important enzymes have emerged. **Restriction endonuclease enzymes** cut DNA where particular base sequences occur. For example, a particular restriction may cut after the occurence of a GGCC sequence. The result of treating DNA with this particular restriction enzyme is the creation of fragments with **sticky ends** as shown here:

ATACCGTAGGCCATTAGATCTGGCCTATGCGA
TATGGCATCCGGTAATCTAGACCGGATACGCT

ATACCGTAGGCC   ATTAGATCTGGCC   TATGCGA
TATGGCAT   CCGGTAATCTAGA   CCGGATACGCT

The ends are "sticky" because they have a tendency to form a base pair with any complementary four base sequence. There are numerous restriction enzymes and they normally occur within cells, because one useful function they can perform is the destruction of foreign DNA, such as that from a virus. **DNA ligase** is an enzyme that seals breaks that occur within a strand in a duplex. This enzyme is thus useful for repairing DNA when breaks occur. Finally, **DNA polymerase** is an enzyme that catalyzes the synthesis of duplex DNA from one strand when a **primer** is used to start the process. How the primer-based synthesis works is shown in the following sequence:

ATACCGTAGGCCATTAGATCTGGCCTATGCGA Single strand
TATGG Primer

+ DNA Polymerase + G + C + A + T →

ATACCGTAGGCCATTAGATCTGGCCTATGCGA
TATGGCATCCGGTAATCTAGACCGGATACGCT

Performing these reactions *in vitro* is the basis for technologies such as gene amplification and crime scene investigation. To illustrate the former, suppose the sequence of at least the start of a gene is known, and it is desired to amplify the amount of DNA that encodes for the gene. Having more DNA would facilitate further study of the gene's properties, possibly through other recombinant DNA efforts where that gene's DNA is inserted into a bacterium or another cell in order to synthesize a large quantity of protein encoded by that gene. A primer can be first synthesized with a **gene machine**. This machine adds the bases A, T, G, and C in the order specified by the user to create short to moderate chains of single strand DNA.

When the primer is added to the single strand DNA, the DNA is single stranded, except where the primer is bound as in the previously shown primer sequence. Allowing DNA polymerase to do its work will extend the primer so that a double strand of DNA is produced. If this duplex is heated, it will fall apart into two single strands, because the thermal energy exceeds the strength of the G-C and A-T associations; this is akin to the "melting" of a material. Primers can again be added and DNA polymerase used to extend the newly added primers according to the instructions provided in the single strand. Overall, the process began with a single strand-primer combination. Each time DNA polymerase is used, the total number of single strands of DNA is doubled. Twenty doublings is a significant gain and leads to a million-fold increase in the amount of DNA originally present. The temperature and polymerization cycling that reproduces DNA in a geometric fashion is known as the **polymerase chain reaction (PCR)**.

PCR and restriction enzymes find practical uses outside life science or biotechnology laboratories. DNA from blood, hair follicles, or other samples from a murder scene can be cut with restriction enzymes to produce a fragment profile that has a high probability of belonging to one person, much like each person has a distinctive finger-print. PCR can be used to increase the amount of DNA obtained from a crime scene sample. Such DNA fingerprinting is powerful because given the size of the total population in an average city, there will be a considerable number of people possessing any particular blood type. Thus, just because the victim's blood type was found on a suspect's clothing, the odds are high that the blood may not belong to the victim; hence, a compelling case requires additional evidence. However, if there is a match between the victim's DNA and that from blood found on the suspect's clothing, then the situation starts to look a worse for the suspect's defense. Conversely, as was found in the early uses of the technology, some convicted suspects were later exonerated by reanalyzing archived crime scene evidence with modern *in vitro* DNA technology.

1.5   Unique vs. Interchangeable Parts Leads to Molecular-Based Classification

A ribosome is composed of different parts that enable mRNA binding and amino acid addition. The parts are called **subunits**. They are characterized and distinguished by how apt they are to settle while being spun in a centrifuge. Based on such physical sorting, they are assigned *S*-values, where *S* stands for a **Svedberg** unit; the larger the value of *S*, the more readily a part settles to the bottom of a centrifuge tube. The unit bears the name of the researcher who studied the behavior of macromolecules and small particles. For his work, Theodor Svedberg (1884–1971), received the Nobel prize in 1926 (http://www.nobel.se/chemistry/laureates/1926/svedberg-bio.html).

One key part of a ribosome is the **16S rRNA** component, which is found in the 30S subunit along with proteins. Many seemingly different cells appear to have interchangeable parts in that the 16S RNAs found in the ribosomal subunits have similar molecular compositions and S-values. Consequently, the genes that encode for the 16S RNAs that arise in seemingly different cells exhibit similar base sequences. When there is a high degree of base overlap in a coding sequence, the DNAs from different sources are said to exhibit high **homology**.

Other cells, however, have been found to possess significantly different components that make up the intact ribosome. The S-value can vary slightly. For example, in eucaryotic cells, the rRNA that corresponds to the 16S rRNA "part" in bacterial cells actually "weighs in" somewhat higher at 18S. More notably, the encoding genes show low homology. Thus, cells are grouped together based on the homology of their 16S rRNA-encoding genes. This means of categorization is akin to lumping Oldsmobiles and Chevrolets together as General Motors products because their fuel pumps are interchangeable. Toyotas, in contrast, have quite different fuel pumps, so they belong in another group of automobile-types. The phylogenic outlook shown in Figure 1.1 is based on 16S rRNA comparisons.

1.6   Cellular Anatomy

Portraits of typical animal and bacterial cells are shown in Figure 1.4. While most cells have similar anatomical details such as a plasma membrane, distinguishing differences exist that can radically alter a cell's properties or capabilities. For example,

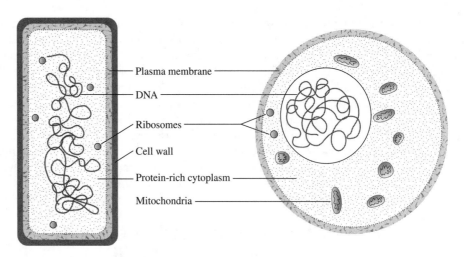

FIGURE 1.4 Some basic features that bacterial and eucaryotic cells share and also differ by. The bacterial cell on the left has a rigid cell wall, flexible plasma membrane, and a protein-rich cytoplasm that also contains DNA and ribosomes. A eucaryotic cell is typically larger than a bacterial cell (not to scale) and it also possesses a plasma membrane and a protein-rich cytoplasm. However, its DNA and energy producing capability are housed in different organelles (nucleus and mitochondria). Eucaryotic cells possess other organelles that are not shown.

a red blood cell and a bacterium both have a plasma membrane that encapsulates all the water and molecules in the cell's interior. However, bacteria have an outer envelope that provides more mechanical strength, thereby allowing bacteria to survive in a greater range of physical and chemical environments. Figure 1.4 also illustrates the compartmentalized nature of the Eucarya.

1.7 | Cellular Physiological Lifestyles

The requirements of a functioning living system will be discussed in more detail in Chapter 3. Here, an outline of the essentials is provided. Whether one considers a bacterium or a moose, there are two invariant requirements: (1) an energy source must be available, and (2) raw materials must exist for building molecules.

Energy is obtained from oxidizing molecules that are imported into the cell. For example, many cells can derive energy from oxidizing a molecule such as glucose. When glucose is oxidized, the reaction is

$$C_6(H_2O)_6 + 6O_2 \rightarrow 6CO_2 + 6H_2O + \text{heat.}$$

Living systems have devised ways to store and direct the use of the energy released from oxidation so that less is immediately lost as heat. Instead, the energy is used to drive the synthesis of proteins and other molecules. Some heat generation still does occur, as, for example, our 98.6 degree Fahrenheit body temperature attests.

The raw materials needed for growth are defined by the elemental composition as well as the amino acid and other special requirements. Often an energy source such as glucose can serve as an energy and raw material (carbon) source. However, sources of nitrogen, phosphorous, and other elements must also exist because a molecule like glucose contains only C, H, and O. Often these other nutrients are found in salts or other dietary intakes. When a cell derives its energy from the oxidation of an organic compound, it is said to have **heterotrophic metabolism**.

Ample cases exist in nature where the energy and carbon source are not the same. An interesting example is provided by the bacteria that enable ore leaching. These bacteria use the energy liberated when iron sulfides (e.g., FeS_2 in pyrite) and other minerals are oxidized and in the process, acid (H^+) is produced. When this microbial reaction is deliberately fostered in a process called **bioleaching**, the acidity tends to mobilize metals such as copper in the ore. However, the reaction can just be a side-effect when excavation occurs. Coal mining provides an example, and a problem can arise. Rainwater and ground water can drain through the mine excavation, and the **acid mine drainage** can negatively impact nearby streams in the watershed. When a cell uses inorganic materials for energy, it is referred to as having **chemolithotropic metabolism.** The cells in our body are heterotrophic.

Regardless of how a cell obtains its energy and raw materials, an efficiency can be defined to quantify how a cell transforms its raw materials to cell mass. Typically, the limiting factor is the basis for defining the efficiency. For example, when all needed materials except glucose are in excess, a **yield** on glucose can be defined as

$$Y = \text{mass cells produced/mass glucose used.}$$

Values tend to be 0.1 to 0.5 g cell water-free/g glucose. Variations arise that reflect the differing efficiencies in using a compound such as glucose for energy and carbon, as well as the nature of the metabolism.

1.8 | Viruses

There is another actor in the biological world that does not fully conform to the genetic, molecular, and physiological attributes we have discussed so far. Viruses are very small particles that are capable of infecting and altering the behavior of bacterial, plant, or animal cells. Without entering a host, a virus cannot reproduce itself. Reproduction occurs when a cell's manufacturing resources are redirected according to the genetic information possessed by the virus. When the virus invades the cell, the instructions for producing the proteins and other molecules that constitute the virus are inserted into the host cell. The host cell then becomes subservient to the virus's instructions. Often the relationship is fatal. After a host cell has produced the components to build many viruses, the components assemble, and they are released to continue the infection by causing the host cell to burst. In other cases, a virus may simply integrate its instructions into a host's genetic material and not produce new virus particles. When infection occurs without reproduction, the condition is **latent.** Many diseases such as chicken pox and AIDS are due to viruses. The fact that a virus invades a host cell contributes to the difficulty in treating viral infections with drugs, because the virus and its lifecycle must be attacked without excessively harming the host cell.

Viruses also carry out some useful functions. In the course of infection and burst, sometimes genetic information from the current host cell is packaged by newly produced viruses. When these viruses invade other cells, some genetic capabilities from the prior host can be passed on to the new host. This is one way in which genetic information is exchanged between cells. If the viral infection is mild or enters a latent state, the new genetic information a host cell receives can be used by the host. Indeed, the use of mutated or mild viral strains is used in the technology of transferring new genetic information to a cell. One example is **gene therapy**, where the goal is to replace a patient's defective gene with a correct one. Gene therapy researchers continue to view viruses as one option for repairing defective genes and remedying genetic-based diseases.

1.9 | Prions

If viruses challenge definitions of what is "alive" versus "dead," then prions test established norms even further. **Prions** are altered proteins that may be capable of fostering their synthesis. Notably, prions are very resistant to physical destruction (e.g., through heat) or to the mechanisms the body uses to recycle materials or to eliminate signals that are no longer valid. Diseases of the brain, such as **Creutzfeldt Jakob disease (humans)**, **chronic wasting disease (CWD)**, and **bovine spongiforme encephalopathy** are linked to the accumulation of prions.

How prions are formed and accumulate in brain tissue is under intense investigation. Among the ideas proposed is that prion proteins are associated with a virus or virus-like particle. Failures to date to find nucleic acids involved have lead to an alternative, "protein-only" hypothesis. The idea is that a protein normally present undergoes an alteration in three-dimensional organization. The altered protein can, however, bind to normal protein. The consequence of the binding encounter is that a normal protein is "refolded" to a degradation-resistant version. A chain reaction can occur that leads to the accumulation of protein in the brain, which, in turn, interferes with normal brain function. More research will reveal if a prion is linked to a disease agent or is itself the culprit.

Because prion-linked diseases affect livestock, wildlife, and humans, many societal stakeholders and researchers from different disciplines are concerned and devoted to elucidating the mechanisms and nature of prion-related diseases. The prospects for prion-related diseases exchanging between interconnected livestock, wildlife, and human populations further amplifies interest. For example, elk, white tail deer, and mule deer have been found to suffer from CWD. Consequently, monitoring studies are underway in Colorado (http://resourcescommittee.house.gov/107cong/forests/2002may16/miller.htm, http://www.nrel.colostate.edu/projects/cwd/) and elsewhere.

EXERCISES |

1.1 (a) If our DNA contained combinations of three bases instead of four, how many amino acids could be encoded when a codon contains one, two, or three bases?

(b) Why do you suppose living systems use four bases instead of three bases in the genetic code?

1.2 If a cell is maintaining 200 proteins and the average number of amino acids per protein is 75, what is the total number of A, G, C, and T bases used to code for the construction of the proteins?

1.3 Match the statement on the left with the best analogous match on the bottom:

A car is either a Toyota or a Ford. _____

The Naval Tomcat aircraft was based on the F14 prototype. _____

Ford water pumps do not work in Toyotas. _____

Each musical measure has the same number of beats. _____

A taped message in *Mission Impossible* incinerates after it is read. _____

(a) Taxonomy (b) 16S RNA gene in Archaea vs. Eucarya (c) mRNA lifetime (d) Phylogeny (e) codon triplet (f) universal genetic code (g) Bill Clinton

1.4 Which compound(s) below would *not* likely be used by a heterotroph for energy?

(a) carbon dioxide   (b) glucose   (c) methane
(d) fructose   (d) reduced iron (Fe°)

1.5 A microbe was found in a fossilized meteor in Antarctica. It appears to have a DNA-like molecule made of 5 different types of base-like molecules. An analysis of other molecules suggests that 25 different amino acid-like molecules make up something that resembles proteins. If there is a genetic code in use, then what is the minimum number of base molecules per codon? Why is the value a minimum?

1.6 If the yield for the growth of *E. coli* on glucose is 0.3 g cell (water-free basis)/g glucose, answer the following:

(a) What will be the mass concentration of *E. coli* on a water-free basis when provided 5 g glucose/liter?

(b) What will be total mass of *E. coli* per liter?

1.7 Assume that a typical protein has 100 amino acids and a "ballpark" molecular weight for an amino acid is 100 g/mol. How many protein molecules are present per 70 kilograms (i.e., average weight of a human) of hydrated animal cells? If a protein's typical dimension is 10 Angstroms ($1 \text{ Å} = 10^{-8}$ cm), could the distance between Pittsburgh and Los Angeles be spanned by aligning the protein molecules end-to-end?

1.8 What type of cell on average has more transcription going on, a growing bacterial cell or a stem cell residing in the bone marrow?

1.9 Where would one most likely look to isolate a new Archaea—in spoiled hamburger or in a boiling hot spring in Yellowstone National Park?

1.10 Identify what is not true about the following statement: *Bacteria and animal cells use the same genetic code (i.e., codons) to store protein "recipes" and use the same ribosomal machinery to translate the information into functional proteins.*

1.11 Assume a cell possesses 1000 genes, but at any point in time, ten percent are being used (i.e., "expressed"). If a typical protein has 100 amino acids, what is the total number of A, G, C, and T bases that encode the cell's (a) expressed and (b) total genetic repertoire?

1.12 A codon in a Martian bacterium contains combinations of four of the five base-like molecules that makes up what passes for Martian DNA. An average bacterium on Earth has 4 thousand genes. On Mars, life is quite different; hence, a larger repertoire of 6 thousand genes is needed to provide flexibility. By what factor is the DNA larger in a Martian bacterium compared to an Earthling bacterium? Assume a Martian protein contains about as many amino acids as an Earthling protein.

(a) 1.2   (b) 1.33   (c) 1.67   (d) 2.0   (e) 2.66

1.13 A new organism may have been found in the University Center dining hall in the cole slaw. What is the best thing to do *first* to characterize the new cell?

(a) Sequence the entire genome.

(b) Measure mRNA stability.

(c) Look for similarity with the 16S RNA encoding stretch of DNA in other known organisms.

System Principles of
Living Systems

Mass Conservation, Cycling, and Kinetics

2.1 | Purpose of This Chapter

The last chapter reviewed cell types, compositions, metabolic lifestyles, and the idea of yield. Raw materials such as glucose are used by cells to maintain and reproduce themselves. Thus, bioengineers often need to assess what mass inputs are required to maintain a living system or to encourage its proliferation. For example, bioengineers design and optimize the large fermentation processes used in the pharmaceutical and food industries to produce cell-derived products such as penicillin and MSG. (Recall Figures 0.1-0.2.) Bioengineers also grow quantities of cells for tissue engineering applications. (Recall Figure 0.4.) Alternately, a bioengineer may be involved with designing a life support system for space flight or patient maintenance. The quantitative assessment of mass inputs and where they go is called **mass balancing**, and an analysis is based on the fact that mass is conserved.

This chapter uses examples to illustrate the basics of mass balancing. Additionally, the subject of **kinetics** is introduced. Kinetics is concerned with how fast a reaction process occurs and what factors affect the rate. When kinetics is used with mass balancing, questions of *"How much?"* and *"How fast?"* can be answered. The examples of exponential and nutrient-limited proliferation of cells are used to introduce the subject of kinetics.

2.2 | Open versus Closed Systems

When accounting for the movement and transformation of mass, a first consideration is to establish whether one is dealing with a **closed** or an **open system**. A **closed system** is one where no exchange of material with the surrounding environment occurs. An example is the food you leave in your refrigerator when you go on vacation. The door remains closed while you are away, so all the reactions and other processes that occur are based on the initial inventory of material and biological agents present in the refrigerator the day you departed. Upon opening the door after your return, you may have had the experience of being impressed by the extensive and multiple transformations that occurred in that closed system.

An **open system** exchanges material with the environment. A lake provides an example. Rivers and rainfall introduce nutrients and water necessary for sustaining plant life and other biota. The lake also absorbs oxygen from the surrounding atmosphere and then emits gases such as carbon dioxide and methane, which originate from the breakdown of organic inputs.

Many experiments are designed to have some facets that emulate a closed system, in order to exercise control over the variables. A cell growth experiment in a flask is an example. The composition of the growth media is defined ahead of time and the system is cut off from other sources of nutrients, so that the effect of a defined set of nutrients can be measured.

Our bodies represent a collection of open systems. On the large scale, we use oxygen from the atmosphere and emit carbon dioxide; hence, we are constantly exchanging mass with the environment. Within the body, mass exchange between organs occurs through the circulatory system and other vehicles.

2.3 | Steady State versus Unsteady State

A process at **steady state** is one where compositions, mass inventories, volumes, and other such variables do not change with time. When variables change with time, the process is termed **unsteady**. An example is provided by water flowing into a sink. When the faucet is turned on and the drain is closed, water flows into the sink causing the level of retained water to rise. If the volume of water is denoted by $V(t)$, then mathematically $dV(t)/dt > 0$. The filling process is unsteady because volume changes with time.

Eventually, the water level rises to the top edge of the sink. The expected result is that water flows out of the sink and onto the floor. A balance is achieved where the rate that water flows into the sink equals the rate at which it exits the sink via overflow. The volume of water is also constant and in this case, the retained volume of water is equal to the volume of the sink. The system is now in a steady state because no variables that describe the system (e.g., volume of water in the sink) are changing with time.

Mass is always conserved, but the approach used to solve balance problems differs depending on whether steady state is achieved or not yet established. For processes that have achieved steady state, mass balances are solved arithmetically and algebraically. Unsteady mass balances typically require the solution of differential equations. Rate information is also needed. This chapter will first consider cases where steady state exists. After discussing kinetics, some examples of solving unsteady-state problems will be provided.

2.4 | Approaches to Performing Mass Balances

The basic steps in performing a mass balance are as follows:

1. Select a basis and draw a diagram showing inputs and outputs.
2. (a) If a limiting material is present, identify it and use the information in formulating and solving the problem.

(b) Search and use a tie component to facilitate a solution or identify the species to balance via an algebraic solution.

The following examples will illustrate what each step means and how to perform it.

2.4.1 Example of basis selection and use: raw material to build a human

How much glucose [$C_6(H_2O)_6$, MW 180] is needed to build a human when glucose provides all the carbon? We start by choosing a basis for the problem. One useful basis is to consider 1 kg of human. Because humans come in different sizes, the calculation on this basis has to be performed only once. The answer can simply be multiplied by the mass of interest to obtain a specific answer.

Basis: 1 kg human, 70 percent water, and 50% carbon on a water-free basis.

Glucose required = 1 kg human * 0.3 kg water-free/kg human * 0.5 kg C/kg
water-free *
180 kg glucose/72 kg C = 0.375 kg glucose(per kg of human).

Note the use of units and their cancellation; explicitly writing the units and tracking the cancellations will help to keep a calculation from going astray. For a 190 pound (86 kg) person, 28.4 pounds of carbon and 32.2 kg of glucose are required. Next time you are in a store, look at the weight of a bag of charcoal and estimate how many bags correspond to your carbon content.

2.4.2 Example of limiting material identification and use: producing cells

What is the maximum cell mass that can be produced from 0.6 g glucose [$C_6(H_2O)_6$, MW 180] and 0.1 g of N as present in NH_4Cl?
    The process diagram is as follows:

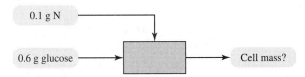

Here again, we start with defining a basis, but whether carbon (C) or nitrogen (N) is the limiting material has to be determined.
    Basis: 1 g cell mass; 70 percent water, and 50 and 14 percent C and N, respectively, on a water-free basis.

C/N needed = 50/14 = 3.57 g C/g N.
C/N available = 0.6 g glucose * 72 g C/180 g glucose/0.1 g N = 2.4 g C/g N.

Because *C/N available* is less than *C/N needed*, there is an excess of N, and C is the limiting material. To compute the maximum cell mass, we assume that all the *limiting* C is used. Then

$$\text{Maximum cell mass} = [\text{g cell water-free/0.5 g C}] * 0.6 \text{ g glucose} *$$
$$(72 \text{ g C/180 g glucose}) = 0.48 \text{ g cell.}$$

This is the *maximum* cell mass, because we examined only the C and N requirements for mass. In practice, some of the carbon in the glucose might be oxidized for energy and released as carbon dioxide. Thus, not all the C in glucose would be available for building cell mass. If glucose oxidation were significant, glucose would actually limit *further* the amount of cell mass produced. (Recall the discussion of yield in the prior chapter.) However, such an upper bound calculation is useful to perform because the range of possibilities (upper and lower) can easily be determined.

2.4.3 Example of algebraic solution approach: setting growth medium composition

The total mass of carbon and nitrogen that can be used to grow cells is 1 kg. Assuming all elemental materials are converted to cell mass at 100 percent, what is the maximum amount of cells that can be produced?

Let the basis be 1 kg total carbon and nitrogen. Also recall that 50 and 14 percent of the cellular dry weight is composed of carbon and nitrogen, respectively. Allow x and y to refer to the mass of carbon and nitrogen used, respectively. There are two unknowns; hence, two independent equations are required to solve the problem:

$$x + y = 1. \tag{2.1a}$$

$$x = (0.5/0.14)y. \tag{2.1b}$$

The first equation states that the total carbon and nitrogen use totals 1 kg. The second equation must be satisfied for no element to be in excess relative to the other. Cellular elemental compositions are used to link carbon to nitrogen use. The second equation cannot be obtained from the first equation by multiplication, adding other equations, or by other manipulation; hence, the two equations are independent.

Solving the two equations results in $x = 0.78$ kg and $y = 0.22$ kg. Thus, when not on a water-free basis, the total amount of cells produced is

$$0.78 \text{ kg C} * [\text{kg cells wet basis/}(0.3 * 0.5 \text{ kg C})] = 5.2 \text{ kg cells.}$$

When one has more than two equations and two unknowns, solution by hand can be cumbersome. Matrix methods are typically used and supported by many software packages. Matrix algebra tools also allow one to double-check that the equations are all independent. Subsequent math and engineering courses will introduce you further to these advanced problem-solving techniques.

A more biologically realistic version of this problem specifies that the total mass of the materials commonly used, which are glucose ($C_6(H_2O)_6$, MW 180) and ammonium chloride (NH_4Cl, MW = 53.5), is limited to 1 kg. Such a constraint might arise on a life science experiment that is to be flown into space. Numerous "science packages" will be flown on the same flight and the total load is fixed;

hence, each experiment's size and/or mass is restricted. If the cell yield (recall Chapter 1) is 0.3 g cell dry weight/g glucose and the masses of glucose and ammonium chloride are x and y, respectively, the algebraic solution is as follows:

Basis: 1 kg total glucose and ammonium chloride; cell 14 percent by weight nitrogen on a water-free basis. Thus,

$$x + y = 1. \tag{2.2a}$$

$$y = (0.3 \text{ g cell/g glucose}) * x * (0.14 \text{ g N/g cell}) *$$
$$53.5 \text{ g NH}_4\text{Cl/14 g N} = 0.16x. \tag{2.2b}$$

Solving the two equations results in $x = 0.86$ kg and $y = 0.14$ kg.

2.4.4 Example of tie component use: using a tracer to measure blood flow in a vessel

Within the blood stream, the concentration of a metabolite is 100 parts per million (1 mg/kg = 1 part per million = 1 ppm). Assume that 1000 mg of metabolite is to be used as a tracer; it is injected over five minutes and the downstream concentration is measured to be 4000 ppm. What is the mass flow of blood in the vessel, assuming good mixing occurs? Estimate the volumetric flow rate of blood based on the answer.

Here, the tracer enters and becomes diluted due to mixing. In the mixing and transport process, the original blood mass is fixed, so it is a tie component. When 100 g total of incoming material is chosen as a basis, the process diagram is that shown by Figure 2.1.

The tracer mass concentrations are on different bases because more mass exits the vessel than entered; hence, one has to be corrected so that the concentrations are on the same basis.

Basis: 100 g *initial total* blood and tracer mass.

Component	Initial Mass Percent	Final Mass Percent
Blood	99.99	99.6
Tracer	0.010	0.400

0.4 g tracer/99.6 g blood * 99.99 g blood/100 g initial

mass = 0.4016 g tracer/100 g initial mass.

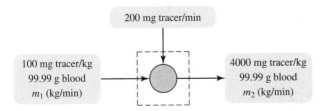

FIGURE 2.1   Process diagram for the injection of a tracer into a blood vessel. The amount of blood that is mixed with the tracer is fixed; hence, blood mass is a tie component. The total mass flow, however, increases somewhat due to the addition of the tracer, which is why the mass flow out (m_2) exceeds the inflow (m_1).

Now the final tracer concentration is on the same basis as the initial tracer concentration. Next calculate the tracer addition as

Concentration change = 0.4016 g tracer/100 g initial mass − 0.0100 g tracer/100 g initial mass = 0.3916 g tracer/100 g initial mass.

Thus the mass flow rate is

100 g initial mass/0.3916 g tracer * 1 g tracer/5 minutes = 51 g/min.

If the density of blood is comparable to water (a slight underestimate), the volumetric flow rate is

51 g/min * 1 liter/1000 g * 60 min/hour = *3.06 liter/hour.*

An algebraic solution can also be found, based on simultaneously solving the total mass and tracer balances around the boundary shown in Figure 2.1:

$$m_2 \text{ [kg/min]} = m_1 \text{ [kg/min]} + 200 * 10^{-6} \text{ kg/min.}$$
$$100 \text{ mg/kg} * m_1 + 200 \text{ mg/min} = 4000 \text{ mg/kg} * (m_1 + 200 * 10^{-6} \text{ kg/min}).$$
$$m_1 = 5.1 * 10^{-2} \text{ kg/min} = 51 \text{ g/min.}$$

Tracer studies are commonly used to find flow rates and to detect blockages. In addition to metabolites, dyes or other easily measured compounds can be employed.

2.5 | Recycle, Bypass, and Purge

When we put an empty soda can into the recycling barrel, our intent is that the aluminum will be melted and used to build a new can. Reusing the can is desirable because it takes one-third less energy to use reduced aluminum to make a can than to start with raw bauxite ore. Thus, less newly mined bauxite and energy are needed, which, when taken together, lessens the environmental impact of putting a new soda can into the market place.

Our bodies and other natural systems also **recycle** raw materials. For example, red blood cells contain iron, which endows a major cellular protein, hemoglobin, with the ability to bind and deliver oxygen to tissues. Red blood cells are replaced about every 120 days and a portion of the iron in them is reused in the replacement cells. Thus, recycling can be a beneficial system attribute and a commonly arising facet of mass balances. The more recycling that occurs, the fewer new raw materials are needed to sustain a system. From a bioengineering perspective, computing how much is needed to sustain cells or a person often entails the consideration of what recycling mechanisms exist.

Bypasses and **purges** are material flows that eliminate excess mass or components from a system. The operation of outdated sewage treatment plants illustrates what "bypassing" means. Storm water from roads and sewage are collected and conveyed through the same piping system to a central wastewater treatment facility. A heavy rain can raise the total flow to the point that the capacity of an old and outdated, outdated treatment plant is exceeded. If all the wastewater rushed in, the tanks in the facility would overflow. To avoid flooding of the facility, a portion of the total input flow is shunted around the plant and directed into the receiving body of water.

Purging is done to eliminate the buildup of some components when recycling is done. For example, reusing aluminum or plastics can result in the accumulation of impurities. Some portion of the recycled material is thus discarded and new material is added to the mix to lower the total impurity level.

The presence of recycling, bypasses, and purges introduces additional complexity to a mass balance analysis. The trick to the analysis is to develop balance relationships that apply over different system boundaries. Specifically, one first accounts for the overall input-output balance akin to what one would do if recycling were not occurring. Then, mass balances are written around a section internal to the process to introduce the effect of the recycling flows.

2.5.1 Recycle and purge example: iron cycling in human body and assessing dietary needs

Analyzing the fate of iron in the body can illustrate how to deal with recycling, purging, and/or bypassing in mass balances. This problem also illustrates how to estimate (1) nutritional requirements and (2) the sensitivity of a living system to a change in the efficiency of a physiological function.

Let us pursue the following question through a mass balance analysis: "*For given values of dietary intake of iron, disposal rate of degenerated red blood cells, and amount of iron from degenerated red blood cells that is recycled, how much ingested iron is eliminated from the body?*" Basically, the question is asking how much of the iron ingested is not used, based on the efficiencies of the cellular- and molecular-based recycle and nutrient absorption mechanisms used by the body.

Balances. The process diagram we assume to model the fate of iron is as follows:

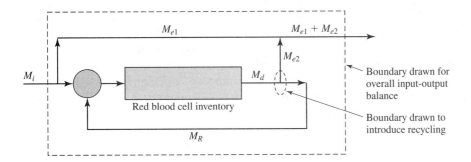

Iron can exist in different forms, such as a salt in food or as bound to protein in a red blood cell. However, we can use iron content information to frame the problem in terms of elemental iron mass flow rates (e.g., mg iron/day). The rates of iron intake, disposal, bypass, purge, and recycle are M_i, M_d, M_{e1}, M_{e2}, and M_R, respectively. To put these rates into the context of the human body, M_{e1} depicts a bypass to an excretory pathway that can eliminate any excess dietary intake from the body. As M_{e1} increases relative to M_i, less iron is "absorbed" from the digestive system. M_R and M_{e2} denote the iron that is recycled from red blood cells or directed to another excretory pathway, respectively.

There are five mass flow rates, and based on the question posed, the values of three (M_i, M_d, and M_R) are known or can be set. We are free to set the value of the dietary intake, M_i. We also know what the rate of red blood cell disposal is, and based on the iron content of red blood cells, the amount of iron available for recycle (M_d) can be determined, as will be shown at the end of the analysis. Finally, we also can specify the value of M_R (or the efficiency of recycle M_R/M_d). Therefore, there are two unknowns, and so two independent equations will be needed to determine the iron mass rates for the set values of M_i, M_d, and M_R.

An overall balance can be written first around the total system boundary (square boundary in the process diagram) as

$$M_i = M_{e1} + M_{e2}. \tag{2.3}$$

A second balance can then be written around the internal boundary shown (circular boundary in the process diagram) to bring recycle (i.e., M_R) into the analysis:

$$M_d = M_{e2} + M_R. \tag{2.4}$$

Using the results to build an operating diagram.   There are now two independent equations that can be used to calculate the unknown values of iron excretion based on specified values of M_i, M_d, and M_R. Specific questions can be now be asked. For example, for the minimum daily iron intake suggested for a typical college age male (8 mg; http://www.cc.nih.gov/ccc/supplements/iron.html), *"What fraction is excreted, assuming that 80 percent of iron in red blood cells is recycled?"* The two balance equations can be used to answer this question. However, it is useful to keep our analysis general, because we may want to ask the same question again, but with a different value for the recycle efficiency. Considering different efficiencies may be of interest, because individual physiologies differ and the effect of such variability on how dietary iron is used may be an important consideration when advising people on dietary needs.

An operating diagram can be constructed based on the balance equations. This diagram can display all the possibilities, and by inspection, the question can be answered for any assumed value of recycle efficiency simply by inspecting the diagram. Defining the recycle fraction to be $\varepsilon = M_R/M_d$, Equations 2.3 and 2.4 can be combined to yield

$$M_{e1}/M_i = 1 - M_d/M_i \,(1 - \varepsilon). \tag{2.5}$$

To proceed, a value is required for M_d. A value can be found by using information on red blood cell inventory, iron content, and lifetime. In adult males, the red blood cells in 100 ml of blood contain approximately 15 g of hemoglobin (Hb). The molecular weight of hemoglobin is 64,500 and each hemoglobin protein contains four iron atoms that are associated with oxygen-binding, heme groups. Thus, for a total of 5 liters of blood, which is typical for an adult male, the iron content is

$$5 \text{ lit blood} * 15 \text{ g Hb}/0.1 \text{ lit} * \text{mol Hb}/64{,}500 \text{ g} * 4 \text{ mol iron/mol Hb} *$$
$$55{,}487 \text{ mg iron/mol iron} = 2581 \text{ mg iron}.$$

Red blood cells are disposed of and replaced every 120 days. This value is an average cell lifetime and in practice, red blood cells are disposed of and replaced daily. Thus, the iron available for recycle is 2581 mg iron/120 = 21.5 mg iron/day.

Using this estimate of how much iron is turning over and an intake of 8 mg/day, we obtain

$$M_{e1}/M_i = 1 - 21.5/8 (1 - \varepsilon) = 1 - 2.67 (1 - \varepsilon) = 2.67\varepsilon - 1.67. \quad (2.6)$$

To answer our original question, it is useful to plot the above equation. Not only will we find what fraction of the iron intake is eliminated when the recycle efficiency is 80 percent, we can in one snapshot see how the system performs and what bounds may exist on efficiency when the system operates under the assumptions we have made. The operating diagram looks like the following figure:

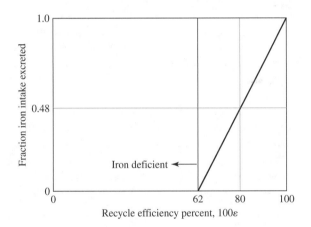

For complete recycling (recycle efficiency percent = 100), all the ingested iron is excreted (i.e., by passed via M_{e1}), which makes sense because there is no need to replenish a loss. If recycling is 80 percent efficient, then roughly half of the ingested iron is excreted. Therefore, for a small drop of only 20 percent in recycle efficiency, ingesting new iron becomes quite important. Interestingly, if the recycle efficiency drops to below 60 percent, the assumptions and calculations indicate that all the ingested iron is used and a person may be on the border of iron deficiency. In practice, about 10 percent of ingested iron is absorbed. From the operating diagram or Equation 2.6, the recycle efficiency is actually quite high (96 percent). More details on iron physiology will be provided in Chapter 9.

2.6 | Kinetics

Why it's useful and definition.   How fast cells can proliferate has a bearing on how rapidly a microbial infection will progress, how fast cell division will repair tissue, and other important events. Mass is transformed in the process of cell proliferation where, for example, the carbon from glucose is incorporated into cell material. Thus, determining how much glucose to add, and over what time span the glucose will be used requires that we know two things. First, how fast the cell population will expand, based on glucose availability must be known. Second, the mass conversion relationship between the amount of glucose used and the amount of cells produced must be established. Thus, being able to predict how fast cells proliferate over time will enable mass balancing and other useful analyses.

Kinetics is the science of specifying the factors influencing a reaction rate and the rate at which a reaction will occur. We can envision that the conversion of raw materials to cells represents an overall reaction. Furthermore, some "laws" may exist that will allow us to quantitatively describe the rate.

Application to describing a cell population increase. Cells duplicate themselves to produce more cells. An extensive literature exists on the analysis and prediction of growth kinetics. Here, we start with one extreme case: the cellular growth rate is under **internal control**. This condition exists when all nutrients are in excess. The limit to how fast cells can proliferate is thus set by how fast the internal machinery can use and transform raw materials and use energy compounds. To provide an analogy, a car can have a full gas tank and a high flow fuel pump. The top speed at which the engine operates, however, may ultimately be limited by how fast combustion and exhaust gas removal occurs within the engine.

One cell divides and becomes two, and then there are four. Such a geometric model works well when all divisions occur at the same time. However, in reality cells divide asynchronously, which means they tend to have an average doubling time, but they do not all divide into two cells at the same moment. Thus, what one observes is a continuous increase in total cell number and total mass versus stepwise (discontinuous) increases. Consequently, a more appropriate description is an exponential expansion. Letting N and t correspond to cell number and time, respectively, the rate that the number of cells increases is proportional to how many cells are present at any one time:

$$dN/dt = \mu_m N. \tag{2.7}$$

The constant μ_m turns the proportionality into an equation and μ_m is referred to as the **maximum specific growth rate**. This description captures the spirit of the geometric increase $(2,4,8,16\ldots)$, but allows for the increase to be continuous in time. Integrating for the initial condition of $N = N_o$ at $t = 0$ yields*

$$N = N_o \exp(\mu_m t). \tag{2.8}$$

The overall time for N to double is

$$t_d = \ln 2/\mu_m = 0.693/\mu_m. \tag{2.9}$$

This equation may look familiar. It resembles the half-life for radioactive decay, which is also described by an exponential process. Akin to each cell doubling at a time distributed around some average, each isotope in a sample will decay with some average time. Individual isotope atoms may decay faster and others slower, but an average value exists. For a given combination of energy source and raw materials, a given cell will exhibit a particular value of μ_m. Thus, the value of μ_m is a useful characteristic for

*Here, Equation (2.7) was "separated" as $dN/N = \mu_m \, dt$, and the both sides were integrated. The use of "separation of variables" will be introduced more formally later for a more complicated problem.

a particular cell-growth environment combination. An upper bound value of μ_m is in the vicinity of 2 h^{-1}, which corresponds to a doubling time of 20 minutes.

When growth is limited by the availability of an energy source or raw material, one would expect that the rate exhibited will be less than μ_m. In terms of the engine analogy used earlier, there is now a limitation outside the engine, such as fuel pump capacity, that limits the engine's speed. The counterpart to the internal control situation is the so-called **first-order dependence** on limiting nutrient concentration. In mathematical terms, the relationship between how fast cells proliferate and limiting nutrient concentration (S) is

$$dN/dt = kNS, \qquad (2.10)$$

where k is a rate constant. Here, doubling S when N is constant will double the proliferation rate; hence, there is a direct linear dependence between rate and the availability of a limiting nutrient. The term "first order" is used because the rate depends on S to the first power. To carry this terminology forward, it would be correct to say that when internal control is the case, the proliferation rate has a **zero-order dependence** on S.

The following figures summarize how the growth kinetics of cells depends on the availability of nutrients. When the nutrient concentration is high, internal control prevails and $\mu = \mu_m$. As nutrients are depleted, external control ($\mu < \mu_m$) sets in. Finally, nutrient exhaustion results in the cessation of growth. Notice how the plots appear when linear versus logarithmic scales are used. On a log scale, the growth curve has a linear segment.

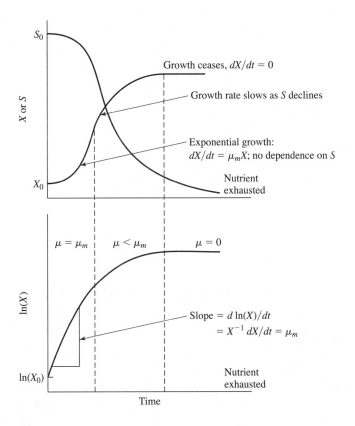

In the next section, we will make use of these kinetic models in unsteady-state mass balances after seeing what is involved with combining rate and conservation relationships in problem solving.

2.7 | Unsteady State Mass Balances

Illustration of analysis with a physical system. To gain a sense of the mathematics associated with analyzing balance problems when steady state is not established, return to the sink filling-overflow image. Referring to the time-dependent volume of water in the sink and constant flow rate of water into the sink as $V(t)$, and Q, respectively, the rate equation is

$$dV(t)/dt = Q. \tag{2.11}$$

Note that the variable is volume, even though the intent is to perform a mass balance. For a fluid like water, volume and mass are proportional; the link is density. Because Q is assumed to be constant, we can rewrite Equation 2.11 as

$$dV(t) = Q\,dt, \text{ where } V = 0 \text{ at } t = 0. \tag{2.12}$$

Integrating both sides of the equation yields

$$V(t) = Q\,t. \tag{2.13}$$

Because $V(t)$ cannot exceed the volume of the sink, V_S, we can calculate the time required to fill the sink as

$$\text{time to fill sink} = V_S/Q. \tag{2.14}$$

After this filling time has elapsed, the process reaches a steady state where the overflow rate equals Q.

This example shows that rate information has to be integrated to describe how a system's mass evolves with time and to answer questions such as *"How long will it take for a given amount of material to accumulate or transform?"*

2.7.1 Biological illustration using mass balancing and kinetics: glucose and time required for a targeted expansion of a cell population when internal control prevails

We can perform a similar analysis, but now in a biological context. For example, we wish to answer the question, *"What is the minimum amount of glucose that should be present at the start of cell growth to support a ten-fold expansion of the cells, and how long will this amount of glucose last?"* One may have to answer such a question when designing a medium to be used to produce a batch of cells for genetic or tissue engineering.

Working with mass concentrations will allow us to perform a mass balance. A good approximation is that the number of cells per volume (N; number/volume) is proportional to the mass concentration of cells (X; mass cells/volume). Thus, we can rewrite Equation 2.8 as

$$X = X_o \exp(\mu_m t). \tag{2.15}$$

The time for a tenfold expansion of total cell mass is given by

$$t = 1/\mu_m \ln 10. \tag{2.16}$$

The above result will answer the second part of the question, but the first part of the question still remains unanswered. Recall that when unsteady problems are encountered, the analysis usually involves combining mass balance and rate information. Thus far, only the rate information has been used.

This system can be viewed as having one initial input and a continuous output with respect to glucose because once set up, no additional inputs of glucose from the environment are available and glucose is consumed to produce cells. Thus, by using the definition of yield (Chapter 1), the fate of glucose can be accounted for as follows: The initial mass of glucose has to be accounted for as (1) unused glucose and (2) the amount transformed to cell material. Allowing the volume of the growth system, initial glucose concentration, and initial cell concentration to be V, S_o, and X_o, respectively, the mass balance on glucose is

$$VS_o = VS(t) + V(X(t) - X_o)/Y.$$

Total initially present = still there + amount used.   (2.17)

The balance in Equation 2.17 resembles how someone estimates how much gas is left in their car's tank when the gas gauge is broken: total gas (gallons) = gas left (gallons) + gas used (gallons) = gas left (gallons) + (miles traveled since filled tank/miles per gallon).

Because volume is constant over time, and noting that we are interested in $X(t)/X_o = 10$, the preceding mass balance can be rewritten as

$$S_o = S(t) + 9X_o/Y. \tag{2.18}$$

The questions can now be answered if some knowledge is applied before inputting some numbers. When $S(t) = 0$, growth will cease. Also as the glucose dwindles toward zero, growth will slow and no longer be exponential. Therefore, the minimum amount of glucose is $9X_o/Y$ and in practice, we may want to add more to insure that the ten-fold expansion target is met in a timely way.

Some reasonable numbers can be used to gain more familiarity with a cell proliferation process. Assume that X_o, Y, and μ_m are 0.1 g/liter, 0.5 g cell/g glucose, and 2 h^{-1}, respectively. The time required to achieve a ten-fold expansion when the growth is exponential is

$$t = 1/2 \ln 10 = 1.15 \text{ h} = 69 \text{ min}. \tag{2.19}$$

The minimum glucose concentration is

$$S_o > 9 * 0.1 \text{ g/lit}/0.5 = 1.8 \text{ g glucose/lit}. \tag{2.20}$$

The medium used for cell growth typically contains 1-10 g glucose/lit, so this answer makes sense. Overall, the calculation illustrates how balances and kinetic information can be used by a bioengineer to plan what resources are needed over time to sustain a living system.

2.7.2 Biological illustration using mass balancing and kinetics: glucose and time required for a targeted expansion of a cell population when external control prevails

In the above example, we assumed that internal control prevailed during the entire time the ten-fold expansion of cells occurred. The time required for external control can likewise be found by combining kinetic and mass balance information. In this case, both X and S vary with time as cell mass concentration increases and a limiting nutrient concentration decreases. The appropriate starting points for the analysis are as follows:

$$dX/dt = kXS \qquad (2.21a)$$

$$S_o = S + (X - X_o)/Y. \qquad (2.21b)$$

Both X and S vary with time, and X depends on S; hence, S and X are *coupled*. This means that neither equation can be solved alone. However, they can be combined by replacing S in Equation 2.21a to yield an equation entirely in terms of X and known constants (X_o, S_o, and Y):

$$dX/dt = kXS = kX[S_o - (X - X_o)Y^{-1}] = -kY^{-1}X^2 + kS_oX + kY^{-1}X_oX. \qquad (2.22)$$

In a sense, we used substitution in a set of simultaneous equations to permit the solution for one variable. Equation 2.22 can be condensed by lumping the constants $-kY^{-1}$ and $(kS_o + kY^{-1}X_o)$ into new constants a and b:

$$dX/dt = aX^2 + bX \qquad (2.23)$$

Finally, the above equation can be solved to obtain an equation for how X depends on time by using the method of **separation of variables**, which will be reintroduced and used again in Chapter 7. This method involves putting similar variables on different sides of the equation to allow for integration:

$$\int_{Xo}^{X} \frac{dX}{aX^2 + bX} = \int_0^t dt = t - 0 = t. \qquad (2.24)$$

The left side of Equation 2.24 can be looked up in any standard calculus book. The answer can be obtained from the standard form

$$\int_{Xo}^{X} \frac{dX}{aX^2 + bX + c} \text{ where } b^2 > 4ac, \qquad (2.25)$$

and it is

$$X(t) = \frac{bX_o\, e^{bt}}{aX_o + b - aX_o\, e^{bt}}. \qquad (2.26)$$

Alternately, $1/(aX^2 + bX) = b^{-1}/X - (ab^{-1})/(aX + b)$, and each smaller integral can be tackled.

A function that relates X to t can thus be established. This result allows for us to predict how X increases with time. Its validity can be checked, in part, by seeing what happens when $t \rightarrow 0$. The result $X = X_o$ should be obtained, and indeed it is. The other limit, $t \rightarrow \infty$, yields $X = X_o + YS_o$, which is in agreement with the convervation of mass.

While the last example has quite a few mathematical details, the overarching aspect to note is in both examples, kinetics and mass balances were combined to answer a question. However, in the prior case of internal control of growth, the kinetic and mass balance portions could be solved independently because growth rate was assumed to not depend on S. In this example, where external control of growth prevailed, the parts were more tightly coupled because growth rate depends on S, and both X and S vary with time. Consequently, the solution to the problem was not as straightforward.

EXERCISES

2.1 A typical cylindrical-shaped bacterial cell is 2 μm long and has a radius of 0.5 μm. Assuming that the yield is 0.3 g cell/g glucose and the density of a hydrated cell is 1.05 g/cm^3, how many molecules of glucose are needed to build one cell?

2.2 A life science experiment is to be flown on a space shuttle mission. The experiment entails studying the effects of microgravity on cell growth kinetics. This experiment is one of many that will be performed, so to manage weight only 1 kg of raw materials can be flown for this particular experiment. What are the optimal amounts of the materials, glucose $(C_6(H_2O)_6)$, ammonium chloride (NH_4Cl), and phosphate salt (KH_2PO_4), that can be brought on the mission if one gram of glucose yields 0.3 g cell on a dry weight basis? Optimal means that no excess mass was flown or occupied space. Notes: glucose (MW = 180, AW C = 12, $C_6(H_2O)_6$), ammonium chloride (MW = 53, AW N = 14, NH_4Cl), and phosphate salt (MW = 136, AW P = 31, KH_2PO_4).

2.3 If 50% of ingested iron is not absorbed by the body and 75% of the iron in red blood cells is recycled, estimate how much iron must be ingested per day to maintain the iron content of the blood. Assume 21.5 mg iron/day is available for recycling.

2.4 A 1,000,000 m^3 lake has a phosphorous content of 5 mg/m^3. A river provides the lake with new water and nutrients. Rainfall provides nutrient-free water. Another river drains the lake. The local precipitation on the lake surface provides 100,000 m^3/year of water.

The flow rate of the outfall is 1,000,000 m^3/year. What must the phosphorous content of the incoming river water be to maintain a constant phosphorous concentration in the lake?

2.5 The surface area of a typical human is 1.8 m^2. If you shed and replace the outer skin cells every 30 days, estimate how many cans of soda are represented by the carbon content of the lost skin cells. Assume that the typical dimension of a skin cell is 50 μm, the density of a hydrated cell is 1 g/cm^3, and a can of soda contains 36 grams of sugar as glucose.

2.6 Average daily water inputs and losses for a human at rest are summarized as follows:

Food	1000 g
Drink	1200 g
Air In	50 g
Air Out	400 g
Metabolic Water Production	300 g
Sweat	350 g (evaporated) + 200 g (damp)
Urine	1400 g
Feces	200 g

If you do not drink any water, will you think that you lost weight and if so, how much?

2.7 Consider the following proposal that has been put to a bioengineer for evaluation. If we put food crops in a space station, then the carbon dioxide released due to the activity of metabolic processes could be used to produce plant material via photosynthesis. Then, once the process was up and going, the astronauts could eat some of the plants and lower the amount of supplemental food they need to bring with them. That is, a steady state inventory of plant mass can be established and the new growth from fixing CO_2 by photosynthesis is harvested. Develop an operating diagram that relates how the fraction of the total glucose respired that can be derived from plants depends on the efficiency of carbon capture by photosynthesis. Assume that an astronaut exhales on average, 100 ml/min of carbon dioxide. Also assume that plants are 70% by weight water, 50% by weight carbon (water-free basis), and 90% of the plant carbon is equivalent to the carbon in glucose-like carbohydrate (i.e., can be metabolized for energy by respiration). Does this proposal seem feasible? Do you have any ideas about how to improve the process?

2.8   The doubling time of an infectious microbe is 20 minutes. How long will it take for the cell population to expand 1000-fold?

2.9   A sink can hold 5 liters of water. When the drain is closed, how long will it take to overflow when the flow from the faucet is 0.50 liter/min?

2.10   A sink can hold 5 liters of water and the drain is closed. You turn on the faucet and then run to answer the phone. The outflow from the faucet is now 0.50 lit/min. After the faucet runs for 8 minutes while you are on the phone, your roommate notices that the faucet is on. The faucet is turned off by your roommate, but due to a worn washer, the faucet drips at 0.0022 lit/min. You both forget about the sink and go to sleep for eight hours. Will there be water on the floor in the morning when you wake up?

2.11   A sink with the drain closed initially holds 5 liters of water ($V_o = 5$ lit). The drain is opened and water begins to drain. The rate that water drains from the sink is proportional to the volume remaining in the sink; hence, $dV(t)/dt = -\alpha V(t)$. Derive an equation that shows how the volume remaining depends on the initial volume (V_o) and α.

2.12   A ten-fold cell proliferation in cell mass occurs. Assume that internal control of the kinetics occurs over the majority of time with $X_o = 0.1$ g/l; $\mu = 2$ h^{-1}; and $Y = 0.5$ g cell/g glucose. Derive how S depends on t and then plot $S(t)$ versus time (t) when S_o is 1.8 g/l and

5 g/l. Discuss how the internal control assumption can cause the long time behavior of $S(t)$ on your plot to differ from reality. What would the long-time section of the curve look like in reality?

2.13   Often a nutrient or energy source is not in excess and the concentration limits the cellular growth rate. For example, some leukemia cells require the amino acid asparagine, and this amino acid is not prevalent in the blood stream. Consequently, one therapy for leukemia is to inject a patient with the enzyme asparaginase, which catalyzes the removal of the amino group, thereby lowering the availability of asparagine to leukemia cells even further. When external control exists on growth kinetics, estimate by what factor the time will increase for a one thousand-fold expansion of cells when the asparagine concentration in the blood stream is reduced from 10^{-7} mol/lit to 10^{-9} mol/lit. Assume that $k = 10^5$ h^{-1} mol^{-1} lit.

2.14   When at rest, a typical person's ventilation rate is 6000 ml/min (in Standard Temperature and Pressure conditions or STP). When one exhales, carbon dioxide is emitted at a rate of 200 ml/min (STP). In a 20 m * 15 m * 8 m lecture hall, there are 50 people.

(a) Estimate the carbon dioxide concentration (mol/lit) in the room after 50 minutes, assuming the concentration was zero at the start of class. Assume that the room is roughly at STP conditions and no one passes out.

(b) Re-estimate the concentration at the end of class, assuming this time that each dimension of the room is cut in half.

(c) Why does the pressure not rise due to all that CO_2 release?

(d) There are places in Africa where lakes emit CO_2. The gas disperses over the surrounding land. Because CO_2 is dense, it tends to sink into the depressions in the surrounding landscape. Small animals fall into these depressions and die from asphyxiation. Larger animals attempt to feed on the smaller animals and they, in turn, perish in the toxic environment. Only a few animals with suitable physiology or stature can survive the high CO_2 concentration that can be established close to the ground, and they prosper from these asphyxiation events by foraging on the victims of asphyxiation. Lakes Nyos and Monoun are particularly infamous for their CO_2 releases, which can also kill humans in surrounding communities. For humans, a CO_2 concentration that exceeds 4.46×10^{-3} mol/lit (10 percent) can be lethal. In view of these facts, what do you conclude about the need for ventilation in the lecture hall?

2.15 A patient has impaired iron elimination. All iron ingested is absorbed (i.e., $M_{e1} = 0$), so feedback control with respect to cellular absorption in the digestive track is defective due to disease.

(a) Show that at steady state, $M_i = M_d (1 - \varepsilon)$.

(b) If M_d still equals 21.5 mg iron/day and the recycle efficiency (ε) is 90 percent, would you advise the patient to continue to ingest 8 mg iron/day?

2.16 Some solution techniques were reviewed/introduced and will be used again. What is the solution to $dy/dt = -ky^2$ for $y = 1$ at $t = 0$? "Solution" means how does y depends on time, t.

2.17 A patient has impaired iron absorption. However, 90 percent of the iron in red blood cells is still recycled and 21.5 mg of iron is available for recycling per day. Generate an operating diagram that depicts how the efficiency of uptake of ingested iron (ε_U; $\varepsilon_U = 1$ means all ingested iron is absorbed rather than eliminated) depends on the daily mass ingested (M_i). Such a diagram could provide disease management guidance.

CHAPTER 3

Requirements and Features of a Functional and Coordinated System

3.1 | Purpose of This Chapter

Engineers differ from biologists and other scientists by their ability to analyze problems and reach goals through science-informed design at the **system level**. Scientists, in contrast, excel at discovering new parts, elucidating function and mechanism, and documenting whether interactions occur between parts. This means that engineers contend with not just the features of the parts, but the desirable attributes gained *and* the challenges presented when all the parts, or a substantial subset, are connected together. The connectivity aspect is the basis for describing work as being performed at the **system level**.

For biological systems, a number of general requirements must be met to allow for the various parts to endow the intact system with attributes such as reproduction, movement, and resilience to environmental changes. Understanding these requirements is essential for bioengineering work at the cellular level or at the human level, where devices or therapies must remedy problems without adversely affecting normal system function.

To expand on the prior chapter, which covered system properties such as overall material needs and kinetics, this chapter shall describe the system-level requirements and features that keep reactions occurring quickly and in the right direction while coordinating their activities. It is important to keep in mind that while the ideas to be presented are vital to bioengineering, they also can apply to other engineering disciplines. Indeed, you may encounter a mechanical engineer who is designing a speed control device to control a car's fuel consumption. That problem may resemble an issue that is pertinent to bioengineering, such as the control of food use by the body.

3.2 | Chemical Reaction Rate Acceleration

When you eat an apple, the carbon and other elements from the apple are transformed into the molecules that constitute your cells. Some of the molecules in the apple (e.g., sugars) are also used to provide you with energy. However, when you leave an apple in the refrigerator, or even in a bowl on the kitchen table, it remains edible for days because little breakdown or deterioration occurs. Obviously, something is going on in your body and cells after you ingest the apple that differs from what occurs when the apple is left on the table.

From the systems standpoint, imagine eating an apple and having it break down as slowly as it does when simply left in the refrigerator. You would need a stomach many times larger that you now have to inventory a large mass of slowly-breaking-down raw materials to allow you to go about your daily activities. Indeed, you would probably have to be built much differently than you now are. Stronger bones and muscles would be required to both support and move this large mass of slowly-used raw materials around along with the rest of you. Perhaps, a point of zero gain would be reached? As your total mass increases, more work will be required to simply walk. Because the rate of raw material supply is low, the point may be reached where the support and operating costs equal or exceed the rate at which resources are made available to the body.

From your experience, that apple is used quickly. The rapid use of resources occurs because the rate at which chemical reactions occur in cells is accelerated by using **catalysts**. Recall from basic chemistry that a catalyst accelerates a chemical reaction without being consumed. Simply, a catalyst is an enabler that gets used over and over with the result that reactions happen faster.

You have encountered catalysts in your daily life. The exhaust system in your car contains catalysts that speed up the oxidation of combustion products. Completely oxidizing these products to carbon dioxide reduces the release of hydrocarbons and other compounds that can contribute to smog formation. In biological systems, special proteins that are called **enzymes** can accelerate the rate of biochemical reactions. The next chapter will provide a molecular view of how enzymes work.

The Cornell biochemist, **James Sumner** established in 1926 that some proteins are enzymes (http://www.nobel.se/chemistry/laureates/1946/). The grandfather of a typical college student was born about that time, so it is amazing how far science and technology have advanced. Sumner isolated an enzyme called urease from jack beans. This enzyme degrades the nitrogen-containing compound, urea, and like most enzymes, the name ends with the suffix, *ase*.

Sumner's demonstration paved the way for the discovery and isolation of hundreds of enzymes. From a more philosophical standpoint, his work established that biological systems are governed by the chemical and physical laws that apply to inanimate systems, as opposed to some special vitalistic force being responsible. You may want to visit http://www.expasy.ch/enzyme/ to gain an appreciation of the vast number of enzymes that have been discovered. In many cases, their structures have been characterized at the atomic level. One way to proceed is to search by putting in the name of a reactant that you know, such as glucose, alcohol, starch, etc.

Returning to the systems view, using enzymes to accelerate reactions presents new advantages and disadvantages. The existence of wins and losses in the performance and operation of a system is often termed **trade-offs** by engineers. Trade-offs are inherent to complex systems and are a useful framework to use for understanding the design of natural systems. They also play a significant role in the design of human-made systems. In design, engineers use creativity, judgment, and social ethics to balance the wins and losses so that a net gain results.

When enzymes are used in biological systems to accelerate the rate at which raw materials are used and transformed, one consequence is that a smaller inventory of raw materials is needed. A simple model can be posed to illustrate the trade-offs. If one apple breaks down linearly over ten days when the rate is not catalyzed, and the daily demand of resources is one apple, then one needs to have ten apples simultaneously breaking down. If catalysis speeds up the rate by a factor of ten, then only one apple a day is needed. Thus, less (apple) storage capacity is required to supply daily needs when catalysts/enzymes are used. However, when enzymes are used, one must procure a new apple each day rather than loading up with ten apples every ten days. Therefore, food procurement becomes a regular daily necessity, which can be viewed as a trade-off against the advantage of needing less storage capacity.

The addition of another system component softens this trade-off. Our bodies store selected resources in concentrated reserves such as fat and glycogen (sugar source), thereby stretching the time needed between apples while not excessively adding to inventory space. Engineers term any component that buffers a system against input upswings or shortfalls as a **capacitance**.

Overall, rates in biological systems are accelerated by the use of enzymes, which function as catalysts. Rate acceleration via the use of catalysts is also commonly used in human-made systems. In any system, the rate at which processes occur can be traded off against the size of the raw material inventory. Adding capacitance can soften the trade-offs.

3.3   Energy Investment to Provide Driving Forces for Nonspontaneous Processes

Spontaneous vs. nonspontaneous.   Bicycle riding is good exercise because you really have to work to crank yourself up a hill. The fun comes when, after you reach the crest, you can coast down the hill and enjoy an exhilarating ride. From the energetic perspective, going up the hill represents a process that requires an investment in energy to occur, because opposite to gravity is not the direction in which movement naturally occurs. In contrast, going downhill is consistent with the direction of gravitational force, so the ride is free. In engineering parlance, a **positive driving force** exists for the downhill ride and the process is **spontaneous**. The uphill ride is **nonspontaneous**.

Chemical reactions, including those that transform raw materials in biological systems, also have spontaneous and nonspontaneous directions. Although gravity does not influence reactions, they nonetheless fall into either the energy investment-requiring or the free-ride category. The subject of **thermodynamics**, which you will study in depth in the upcoming semesters, quantifies whether an investment is needed or not and what factors add or detract from reaction spontaneity.

One type of biological reaction that is not spontaneous is the synthesis of proteins from amino acids. Here, small molecules are connected together to form a very large molecule. **Entropy**, the tendency for disorder to prevail, favors the amino acids remaining unconnected and randomly mixing in the medium such that no particular organization forms and they are not forced to be in proximity to each other. To overcome the entropic tendency, an investment of energy is required. One can thus view biological systems as consumers of energy to combat entropy. Such an idea has been noted by many, including the famous physicist, **Schrödinger**, who coined the word **negentropy** to emphasize the effort needed to combat the tendency for disorder in complex biological systems. This view conveys a key idea that living systems are not immune to the laws of chemistry and physics. However, other attributes are also important, such as the control issues discussed at the end of this chapter.

Coupling pushes nonspontaneous processes.   When a targeted investment of energy is needed, biological systems use a strategy termed **coupling**. By linking a nonspontaneous reaction to a highly spontaneous reaction, the net process can go forward. A mechanical analogy can be made with respect to our bicycle riding experience. The energy released from some other activity can be stored in a battery. Upon release, the energy stored in a battery can then be used to drive a motor that pushes a bicycle and rider up a hill much like a jet is catapulted off the deck of an aircraft carrier. If the stored energy was large enough, the hill's crest can be reached so that the next spontaneous process (downhill ride) can occur.

Be careful, however, when doing the accounting associated with coupling. The energy released from going down a hill can be stored and then recovered to assist with the ascent of the next hill. However, losses will occur through friction, resulting in the production of heat. Thus, perpetual motion, whereby after an initial input of energy you cycle up and down hills without any additional work, is not possible. Rather, some work must be done constantly to keep the coupling mechanisms replenished and ready for use.

ATP coupling in the reactions that occur in biological systems.   In biological systems, the "battery" that stores energy is represented by the bonds in the molecule **Adenosine triphosphate** (abbreviated as **ATP**); the molecule is shown in Figure 3.1. Some surplus energy from spontaneous reactions is stored by adding a new "high-energy" phosphate–phosphate bond to **Adenosine diphosphate** (**ADP**), which thus converts ADP to ATP. As Figure 3.1 shows, the reverse process ATP $\rightarrow$ ADP release the stored energy. The energy stored in ATP is drawn upon to push reactions forward that would not normally occur such as protein synthesis. How such a reaction couple works is shown in Figure 3.2.

The energy stored in each phosphate–phosphate bond in ATP is about 7000 cal/mole. Recall that a **calorie** is the amount of energy that, if converted entirely to heat, would raise the temperature of one milliliter of water by one degree Centigrade. The concentration of ATP in a cell is roughly 1×10^{-3} moles/liter. If all the stored energy was simultaneously released and converted to heat, one milliliter of water would heat up by much less than one degree Centigrade. You may think that this represents very little energy, and you are right. It does not mean, however, that biological systems do not harness and consume much energy. What must be considered is that while the ATP inventory is low, the turnover is very high. At any point in time, many ATP molecules are being produced, thereby storing some of the energy released by some reactions, while simultaneously, many ATP molecules are also being used to push nonspontaneous reactions. In a sense, the energetic activity is akin to a marginally successful yet active gambler's checking account. The deposit and withdrawal activity is large and frequent, but the fixed balance is always small.

To summarize, some reactions proceed readily, whereas others require an energy input. Entropy conspires to defeat molecular organization, which contributes to

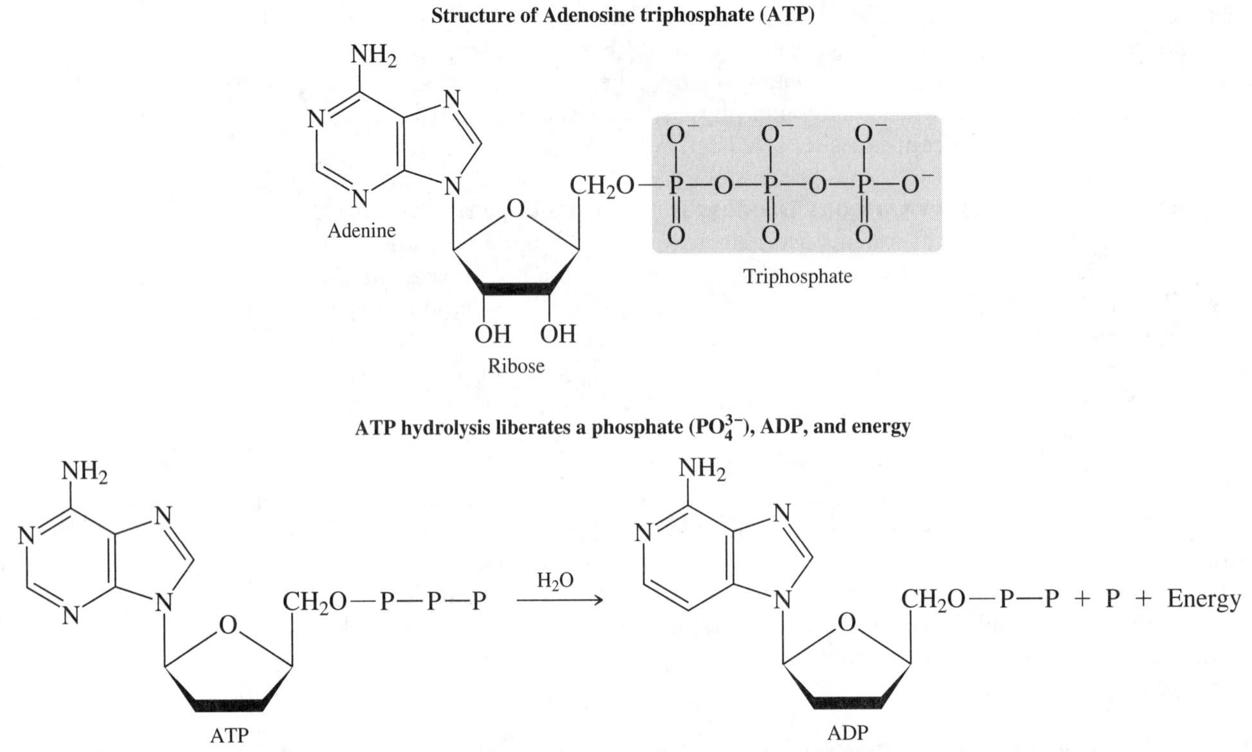

Structure of Adenosine triphosphate (ATP)

ATP hydrolysis liberates a phosphate (PO$_4^{3-}$), ADP, and energy

FIGURE 3.1   Structure of ATP and the hydrolysis reaction that liberates a phosphate from ATP to yield ADP and the energy that was stored in the phosphate–phosphate bond. To simplify the depiction of the hydrolysis reaction, the skeleton of ATP is shown and phosphate is abbreviated as "P."

making a process nonspontaneous. The strategy of coupling a nonspontaneous process to a spontaneous one allows for a sequential set of processes to proceed without getting stopped by a hill. Coupling is a common feature in the operation of biological and human-made systems.

3.4 | Control and Communication Systems

We have established that the many reactions that occur within cells are accelerated by enzymes. Also, the energetic "economy" is vibrant: large deposits and withdrawals constantly occur. Biological systems are thus endowed with speed, and the net directionality of reactions toward organization overcomes entropy.

A vital overarching question is *"How is all this varied and accelerated activity coordinated so that it all works together coherently and efficiently?"* To be more specific, *"What insures that a particular set of reactions does not consume an excessive amount of ATP or overproduce a particular amino acid?"* *"If an enzyme is unnecessary, how can its presence be eliminated from the cell so resources can be used to build other more useful enzymes?"*

Such questions on control and coordination must also be addressed in other complex systems, such as chemical plants and electrical power generation and supply

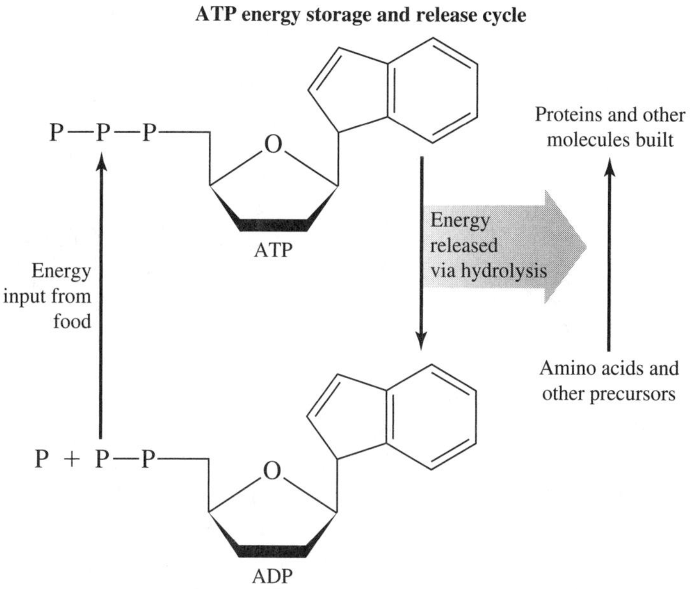

ATP energy storage and release cycle

FIGURE 3.2 The interconversion of ATP and ADP and energy coupling. The hydrolysis of the terminal phosphate of ATP yields on the order of 7 kcal/mole of energy. This energy can be used to power the synthesis of proteins and other molecules. The formation of ATP from ADP and phosphate (P) reverses the hydrolysis reaction, and reversal requires a new input of energy.

networks. Interestingly, many of the control and coordination strategies used by biological systems are emulated in human-built systems. Therefore, understanding the strategies will complete the overview of biological systems and provide a foundation that is shared by several engineering disciplines.

Types of control systems. All control systems somehow measure an output of a process, and in response to the value of the measurement, some aspect of the process may be altered. There are two basic types of control systems: (1) **logical** and (2) **continuous variable**. Logical variable control systems are sometimes referred to as **ON-OFF** or **discrete variable** systems. The appliances and devices we encounter daily can illustrate the nature of these two types of control systems.

An example of a logical-type control system is a room air conditioner. A temperature-measuring device continually samples the room temperature. If the thermostat is set to 70 degrees Fahrenheit and the room temperature remains at less than that value, the air conditioner is in an OFF state. If the temperature exceeds 70 degrees Fahrenheit, the air conditioner is turned on and remains on as long as the room temperature exceeds the threshold value. In this control system, only the **states** matter. It does not matter if the temperature is 50 or 68 degrees Fahrenheit; the state of the measured variable is less than the threshold value, so the air conditioner will remain off. It also does not matter if the temperature is 72 or 90 degrees Fahrenheit; in both cases, the state is such that the temperature (measured variable) is greater than the threshold value, so the air conditioner will be turned on and operate at the same rate.

How you react to managing the temperature inside your car upon entry can illustrate how a continuous-variable feedback control system operates. Let's assume you wish to maintain the car's interior temperature (T) at 70 degrees Fahrenheit. Such a target is called a **set point** in control system terminology. As you enter the car after shopping on a mild spring day, you sense the temperature in the passenger compartment is 80 degrees Fahrenheit. To cool the car's interior, you turn on the air conditioner and choose a low air blower setting to start cooling. If instead it was a very hot summer day, and when you entered the car you sensed the temperature to be a sizzling 100 degrees Fahrenheit in the front seat, you would likely chose a very high blower setting to maximize the cooling rate so that your discomfort time would be minimized. What you did was link the cooling to the amount of difference between the car's interior temperature and the desired target, 70 degrees Fahrenheit. That is, unlike the ON/OFF control strategy, the cooling system's response was increased as the deviation from the target increased.

The amount of change in the control system's response (e.g., air conditioner blower output) per unit deviation from the target is often called the **control system's gain**. A **high gain system** is one that is very responsive to small deviations from the set point. Also note that in this case, the control system's response is against the deviation. The system responds to a positive value of the deviation, $T - 70$, by an action that reduces the value of the deviation. When a control system's action opposes changes, it is termed a **negative feedback control system**.

In the winter, when the temperature is below the set point of 70 degrees Fahrenheit, the same control strategy can be applied, but now with the heater operating. The colder the winter day, the higher the blower setting used to transmit warm air from the heater to the passenger compartment. As with the cooling example, the farther away from the set point, the greater the heating rate chosen. Also, the action taken counters the deviation from the set point; the blower speed increases as $T - 70$ decreases, which, again, is the nature of a negative feedback system.

Figure 3.3 shows how an engineer may design and represent such a negative feedback control system. The aim of the system shown in Figure 3.3 is to control the concentration and flow rate of a fluid. Imagine that a concentrated syrup is being

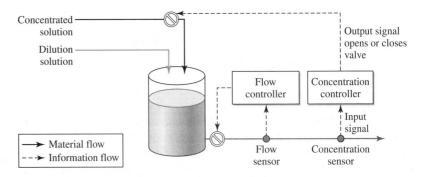

FIGURE 3.3 An example of a negative-feedback, continuous-control system used in a human-engineered system. The objective is to dilute a concentrate such that the flow rate and exit concentration are kept as constant as possible. A sensor measures flow rate. If the error from the set point is negative (actual rate – desired rate), the flow controller opens a valve to an extent that will increase the flow rate such that the target value is hit. The concentration is measured as well. If too dilute, the error signal processed by the concentration controller will result in increased inflow of concentrate.

blended with water, and the aim is to have a constant flow rate and fixed sugar concentration exiting from the blending operation.

A biological counterpart to the human-made control system is shown in Figure 3.4. While the details of the chemistry may look complex, the control logic can be understood. One of the tasks the reactions perform is to degrade and oxidize glucose, and in the process store some of the energy originally possessed by glucose in ATP. Note that two reactions release enough energy to drive the formation of ATP from ADP and in the process, some energy is stored to drive a nonspontaneous

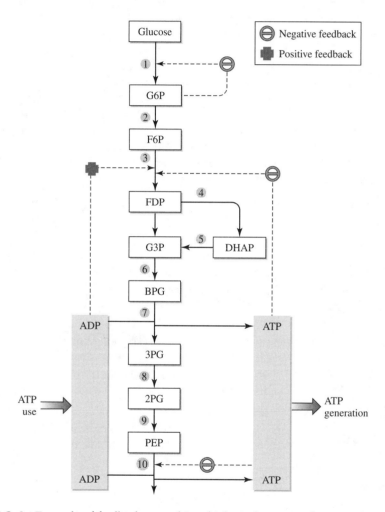

FIGURE 3.4 Example of feedback control in a biological system. Glucose is degraded by a series of enzymatic reactions and reactions 7 and 10 produce ATP. Because ATP production by glucose metabolism and ATP use via biosynthesis need to be kept within a reasonable balance, feedback control mechanisms are employed. If the ATP concentration is too low (demand exceeds supply) or too high (demand drops relative to supply), the rates of the reactions that produce ATP are altered via the ATP or ADP "signals." A mix of negative (excess ATP slows a reaction) and positive (low ATP and thus high ADP stimulates a reaction) feedback is used. The abbreviations are: G6P, glucose-6-phosphate; F6P, fructose-6-phosphate; G3P, glyceraldehydes-3-phosphate; DHAP, dihydroxyacetone phosphate; BPG, biphosphoglycerate; 3PG, 3-phosphoglycerate; 2PG, 2-phosphoglycerate, and PEP, phosphoenolpyruvate.

reaction such as protein synthesis. Thus, if too little or too much ATP is produced relative to demand, feedback controls are desirable for either accelerating or slowing down the enzyme-catalyzed reactions. Figure 3.4 shows that two reactions (3 and 10) are subject to feedback control by ATP.

Overall, in biological systems, ON/OFF and continuous variable control systems operate at the gene expression, metabolic, cellular, and other levels. One outcome is that we tend to have fairly uniform body temperature, red blood cell concentrations, metabolite concentrations, etc. The outcome of steady and bounded operation is often called **homeostasis**, a termed coined by the American physiologist **Walter Cannon** (1871–1945, http://www.encyclopedia.com/articles/02251.html).

While homeostasis is an important concept and objective for many biological control systems, one needs to keep in mind that the scope of control systems is not limited to maintaining some status quo. Indeed, we all originated from a fertilized cell, so control systems also manage the temporal development and construction of biological systems in a precise way. Lastly, as you can imagine, when these control systems are not correctly built or become overtaxed, disease or metabolic disorders such as diabetes can result. Some examples of bioengineering endeavors are developing tools to uncover control system problems, improving patient monitoring devices (e.g., blood glucose measurement kits for diabetics), and helping to produce drugs that compensate or otherwise address the deleterious consequences of improperly functioning control systems.

EXERCISES

3.1 You are hiking in the mountains. You estimate that you will need at least 6 liters (6 kg) of water a day to sustain yourself. You have two choices: (1) carry a 1-liter bottle (1 kg) and fill up periodically when water is nearby or (2) carry 6 liters (6 kg). What are the trade-offs associated with these different solutions? Note that a typical backpacker carries about 18 kg of food, shelter, etc., so the water carried adds to this weight.

3.2 You are contemplating the replacement of your current wood stove. It has cracks that admit a lot of air. Consequently, airflow is high. The high airflow ignites wood quickly and burns it "fast and furious". The new wood stove has lower airflow. What are the trade-offs associated with replacing the old wood stove?

3.3 A net decrease in energy is required for a process to be spontaneous. If a biological reaction has an energy change of 10 kcal/mol, how much ATP has to be hydrolyzed to make the reaction spontaneous when coupled to ATP hydrolysis?

3.4 A one kg mass falls one meter. When it falls, it compresses a spring. After the spring compresses, it extends and pushes the mass back up in the air. During compression, 10 percent of the original energy is lost. How much supplemental energy is required to push the mass back up to the original height of one meter?

3.5 A one cubic centimeter volume of cells can produce 0.03 moles of ATP from an energy source such as glucose. This ATP is used to drive biosynthesis. The concentration of ATP in a cell is also typically 10^{-3} mol/liter.

(a) If all the ATP produced was not used to drive biosynthesis and other processes, and instead rapidly broke down to ADP, what would be the temperature change of the cells?

(b) How much energy is being transacted compared to that stored as free ATP?

3.6 A metabolic reaction occurs at a rate of 10^{-2} mol/h g cell. This means that every hour, within one gram of cells, 10^{-2} moles of some metabolite is enzymatically transformed to another molecule. The energy released per mole of reacted metabolite is 5 kcal.

(a) How much ATP can be generated from this reaction per gram of cell in one hour, as opposed to the released energy appearing as heat?

(b) If the energy was not captured and stored in ATP, how much would one gram of cells heat up 100 cubic centimeters (100 g) of water in fifty minutes?

3.7 You are contemplating increasing the airflow through your wood stove by installing different exhaust piping. Currently, you believe that the airflow is too low because the logs burn slowly and the heat released to the room is on the low side of tolerable. In fact, you now put a fair number of logs in the stove in order to achieve a barely tolerable level of heating.

(a) If you increase airflow and load the same amount of wood, what could be an adverse consequence?

(b) What are two control strategies you could use to make the stove with new exhaust piping provide the desired level of room heating?

3.8 Classify the following as continuous or discrete-variable control systems and explain your answer.

(a) Room air conditioner with a constant-speed fan

(b) Shivering that becomes more intense as the core body temperature drops

(c) The thermostat in a car's cooling system that admits water to the radiator when the water temperature exceeds 220 degrees Fahrenheit.

3.9 Figure 3.4 shows, in part, how glucose is degraded and the reactions involved, where ATP formation can store the energy released by oxidative reactions. The early steps basically entail converting glucose to another sugar, fructose. Glucose and fructose also acquire phosphate groups (glucose-6-phosphate, G6P; fructose-6-phosphate; F6P), which endow them with a negative charge that prevents them from leaking out of the cell. The rest of the steps break the six carbon sugars down to three carbon compounds and then oxidize the glucose-derived fragments.

(a) Will a cell with high biosynthetic demand for ATP and low ATP production rate have a high or low value of ATP relative to ADP?

(b) How will a high value of ADP concentration relative to ATP concentration influence the rate of glucose use? Answer this question in terms of (i) which reaction rates are affected, and (ii) what the net effect is on the rate of glucose metabolism.

(c) How will a high value of ATP concentration relative to ADP concentration affect the rate of glucose use?

3.10 The autopilot on a boat works as follows: A desired course heading is inputted (e.g., due east). The global positioning system continually monitors the boat's position. Over some time span, the boat's actual trajectory is computed (e.g., NE). If necessary, the boat's steering is adjusted to reacquire the desired course.

(a) Would a discrete or continuous control system be used?

(b) What is the signal the controller uses if the desired direction of travel is eastward?

(c) A boat moving at 10 knots is exposed to a 1 knot cross current. The positioning system is accurate to 10 meters. Based on this phenomenon, what are the trade-offs associated with computing the error signal every 10 seconds or 1 hour? What seems to be a more reasonable time scale based on the trade-offs? (Note: A "knot" is one nautical mile per hour.)

(d) For every degree the boat is off course, the rudder can be set to rotate 5 or 20 degrees. Which setting corresponds to a high-gain control system and what are some disadvantages of using high gain control?

CHAPTER 4

Bioenergetics

4.1 | Purpose of This Chapter

In Chapter 3, the requirements for a functional and controlled system were presented. Among the important requirements described was the production and coupled use of energy. Thus, to complement Chapter 2, which covered how to determine the material requirements of living systems in terms of concrete numbers, this chapter covers the quantitative aspects of bioenergetics. You will learn how to estimate energy use and needs as well as more details on the mechanisms that are used to produce and store energy in cells. Such knowledge further enables bioengineers to design systems that support living systems or perform other tasks. The next two chapters will elaborate further on the mechanisms and analysis of the other important requirements and features: catalyzed rate acceleration and control systems.

An overview of thermodynamics, which has applications to every engineering and scientific discipline, is first presented. We shall limit ourselves to the **First Law** and its applications to living systems. The large system viewpoint is first taken where the details of what each cell does in an intact system are lumped together. Thereafter, the cellular mechanisms that are used to produce and store energy are discussed and quantified. Applications of this knowledge are presented throughout this chapter.

4.2 | Units in Bioenergetics

Recall that a **calorie** is defined as the amount of energy that will raise one gram of water one degree Centigrade. One **millicalorie** (mcal) and one **kilocalorie** (kcal) are 10^{-3} and 10^3 calories, respectively. While a variety of other units are used for work and energy such as Joules, we will stick to the calorie unit because its definition is more amenable to visualization, and because it is still commonly used in conveying dietary information on food labels and elsewhere in daily life.

One point of confusion can arise even when we stick to calories as an accounting unit. Food "calories" listed on labels and dietary publications often really refer to kilocalories (thousands of calories). Sometimes they are called "big" calories and referred to as "Calories" as opposed to "calories." We will be precise here and specifically state whether "calories" or "kilocalories" are meant, but be aware of the practice of dropping the "kilo" prefix in some fields.

4.3 | Sensible versus Latent Heat

When a calorie of energy is put into a gram of water, the temperature rises. The addition of more heat will continue to raise the water's temperature until the boiling point is reached. At this point, the temperature rise will cease until all the water has been vaporized. Initially, the heat increases the motion of the water molecules. At the boiling point, the remaining intermolecular attractions between water molecules are being broken. When heat input results in a temperature change, it is referred to as **sensible heat**. When no apparent temperature change occurs, it is called a **latent heat** effect. The melting of a solid and liquid vaporization are typical phenomena that exhibit a latent heat effect.

The heat capacity of a material measures the sensible heat effect. For water, the heat capacity is 1 degree Centigrade/g calorie. When vaporization is involved, the latent heat equals the heat of vaporization. The latent heat of water is 10 kcal/mole (555.56 cal/gram). Water is atypical in that its heat capacity and heat of vaporization are both high. This means that water can absorb a lot of heat and resist temperature change or vaporization.

4.4   The First Law of Thermodynamics Works on All Scales

Thermodynamics is a subject that is concerned with the accounting of energy and its fate. Systems ranging from automobile engines to a moving amoeba are governed by the Laws of Thermodynamics. The **First Law of Thermodynamics** states that

Energy is neither created nor destroyed; it is conserved.

You may have already encountered this law in your physics class. A classic example used to illustrate energy conservation is a spring with a weight attached. Work is done to extend the spring. Upon release, the weight bobs up and down as the spring compresses and extends. The total energy equals the initial work done. The energy, however, is partitioned between kinetic (weight in motion) and potential (spring compression and extension) components.

4.5 | Using the First Law in Energy Balancing

The First Law can be written in equation form to enable calculations and predictions. Denoting E, W, and Q as the total energy, work performed, and heat released, conservation of energy requires that

$$E = W + Q. \tag{4.1}$$

The time derivative of both sides of Equation 4.1 yields the rate at which energy is used, namely,

$$dE/dt = dW/dt + dQ/dt = P + q, \tag{4.2}$$

which equals the sum of power (P, rate of work) and the rate of heat release (q).

In any accounting system, a sign convention is needed to distinguish a debit from a deposit. W and Q will be regarded as positive if the system performs work and heat is released to the environment. Let us now illustrate the accounting system and gain a sense of the magnitude of some values for a human.

4.6   Representative Values in Bioenergetics at the Human Scale

Basal metabolic rate of a human.   When a 70 kg student is sitting and reading this book as opposed to running to class, the heat release rate is about 72 kcal/hour (i.e., $q = 72$ kcal/h). This measurement corresponds to conditions where minimal external work is being performed, so the value reflects a baseline, minimal rate. Accordingly, 72 kcal/hour heat release is called a **basal metabolic rate**.

What is the origin of the 72 kcal/hour basal rate? When food is "burned", about 60% is quickly released as heat. Some energy from food is also captured in the form of ATP, and then used to "power" nonspontaneous chemical reactions, heart cells, and other consumptive processes. When ATP and other reserves are used to power nonspontaneous processes, there can also be heat release, because ATP hydrolysis often provides more energy than required to insure the process occurs in a particular direction. Additionally, mechanical processes all have friction, which manifests as heat. Thus, the basal rate of 72 kcal/hour is often regarded to be a reasonable estimate of a person's minimal energy needs for survival.

Putting these facts together, and given that there are 24 hours per day, indicates that food intake should provide about 2,000 kcal/day. The actual value someone needs to maintain body mass will depend on size, because the 72 kcal/hour "average" is based on a 70 kg person. Additionally, the proportion of your mass that is made up of muscle and other tissue can raise or lower your basal metabolic rate because, each tissue differs in its metabolic rate. Food labels are based on an average 2000 kcal/day requirement. Recall that "calories" on the label actually refer to kilocalories and 2000 is a round number that takes into account the range of people's sizes and their lifestyles.

Heat is typically a waste product of energy conversion processes. For example, friction in the axles of a bicycle results in heat and the incomplete conversion of the rider's cranking work into propelling the rider and bicycle forward. However, in the case of mammals such as humans, the "waste" heat is useful because it maintains the operating temperature at 98.6 degrees Fahrenheit even when it is winter. A constant and elevated temperature is desirable because chemical reactions occur faster as temperature increases. However, heat production can be a problem in the summer, as anyone who has suffered from heat exhaustion can attest.

As Chapter 3 introduced, homeostatic mechanisms attempt to retain or dissipate waste heat in order to maintain the body temperature within a controlled, narrow range. Heat retention is accomplished by restricting blood vessels near the skin surface; exposing less warm blood to a cold boundary lowers the rate at which heat can escape. Heat dissipation occurs by the opening of blood vessels and using the high latent heat of vaporization of water to soak up heat in the process of sweating.

Performing work beyond the basal amount and equivalents.   Running to class will require more energy use because additional work is being performed. We can estimate such extra energy use. As a bright engineering student, you can perform other energy balance calculations to scrutinize the claims of exercise and diet info-mercials and web sites.

To provide an example, let's assume your class is on the tenth floor of a building. You enter the building and find that the elevator is unavailable and not arriving to convey you to class. This is not normally a major problem, but today an exam is scheduled and you want to arrive on time. Reluctantly, you hit the stairs and climb. What is your rate of energy consumption and how does it compare to the basal metabolic rate?

The extra energy used, which we denote by e, equals the work you perform by climbing the stairs. This work, in turn, raises your potential energy in a gravitational field. Thus, in terms of m, g, and h, which are your mass, gravitational acceleration (9.8 m/s^2), and height ascended, respectively, the extra work is

$$e = \text{work performed} = \text{gain in potential energy} = mgh. \qquad (4.3)$$

The *rate* of extra energy use when m and g are constant is

$$de/dt = d(mgh)/dt = mg\,dh/dt. \qquad (4.4)$$

This result makes sense; the more massive you are (m) and/or the faster you climb (dh/dt), the more rapidly energy must be used (de/dt). If you weigh 70 kg, each floor spans 5 m, and you ascend at a steady rate over 5 minutes (i.e., 0.6 km/h), then

$$de/dt = 70 \text{ kg} * 9.8 \text{ m/s}^2 * 10 * 5 \text{ m/5 min} * \text{min/60 s} = 114.33 \text{ Joule/s.} \quad (4.5)$$

There are 4.2 Joules per calorie; hence, the extra work is about 100 kcal/h. While significant, the value of e is in the ballpark of the basal metabolic rate.

Another way to look at the result is to compute how much total energy is needed over that five minutes if you just rested or, instead, ascended the stairs:

Basal use for 5 minutes = 72 kcal/h * 5/60 h = 6.0 kcal.
Required to ascend stairs = 100 kcal/h * 5/60 h = 8.3 kcal.
Total energy use when ascending stairs = 5.85 + 8.3 = 14.3 kcal.

Thus, over five minutes you have more than doubled your rate energy use when you ascend the stairs to class versus remaining in a resting state. The amount of increase is actually greater because other energy consuming processes were not accounted for.

Such results are also often expressed in **equivalents** to add more meaning to a numerical answer. Using equivalents and comparisons puts calculations in an informative perspective. For example, one pound of fat contains roughly 3500 kcal. Thus, the stair climbing work over five minutes is equivalent to $8.3/3500 = 0.00237$ pounds of fat. Overall, we find that (1) our basal metabolic rate is significant compared to a physical activity, (2) body fat can store a lot of energy in terms of the demands of physical activities, and (3) when food is ingested at a pace that does not match energy use, an imbalance can easily occur.

4.7   How Energy Is Produced, Stored, and Transduced at the Cellular Level

In Chapter 3, the ATP molecule was introduced. Recall that the bonds in ATP store energy from oxidative reactions. This energy is then released when and where needed in order to drive nonspontaneous reactions, in a process called coupling. Heat is released by biological systems because each overall reaction that captures energy (ATP formation) or uses energy (coupled to ATP hydrolysis) is less than 100 percent efficient. Even when each energy-capturing reaction is highly efficient, the slight inefficiency of the multitude of reactions occurring simultaneously can add up to significant heat production.

Substrate level ATP formation.   There are two ways in which cells capture energy and store energy in the ATP molecule. The first is called the **substrate level ATP formation** and it was introduced in Chapter 3. As the name suggests, the capturing and storage of energy occurs at the level of individual biochemical reactants (i.e., enzyme substrates) and reactions. To illustrate, assume that metabolite A reacts to form B and heat is produced; that is,

$$A \rightarrow B + \text{Heat}. \tag{4.6}$$

This reaction can be coupled to ATP formation to reduce the amount of waste heat by storing the some of the energy released in the bonds of the ATP molecule:

$$A + ADP + Pi \rightarrow B + ATP + \text{less heat}. \tag{4.7}$$

Energy storage uses chemical concentration differences across the membrane.   The other vehicle is known as **oxidative level ATP formation**. The main difference is that the energy released by a reaction is first stored by an **energized cell membrane**. This repository is then drawn upon to help produce ATP when ATP is needed to propel nonspontaneous processes. Thus, there is an intermediary storage depot and the energy released from a large number of reactions can be stored in the same repository, much like a neighborhood gas station fuels all the local cars.

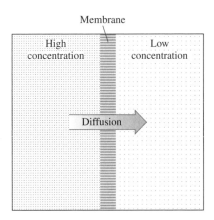

FIGURE 4.1 Illustration of diffusion. The higher concentration of solute on the left drives movement from left to right across the membrane until the concentrations on both sides are equal.

You are probably wondering how a cell membrane can be energized. The researcher **Peter Mitchell** pondered the same question and developed what is now known as the **chemiosmotic theory**, which earned him the Nobel prize in chemistry in 1978 (http://www.nobel.se/chemistry/laureates/1978/). To gain a sense of how membranes can store energy, recall the process of diffusion. Consider two sugar solutions separated by a membrane, as shown in Figure 4.1. If the sugar concentration on the left is initially higher than on the right, then as time proceeds, the sugar will diffuse from left to right until the right- and left-side concentrations are equal, as long as the sugar molecules can fit through the membrane's pores. What will occur is the movement to equilibrium via an energetically spontaneous process. This process is akin to a ball falling to the earth's surface in response to gravity. Work will be performed by ball when it hits the ground that is equal to the ball's initial potential energy.

If we now start with equal sugar concentrations on both sides of the membrane, how can the left side be made to have a higher sugar concentration? One has to run the diffusion experiment backwards from the energetic standpoint. Rather than having a spontaneous process that releases potential energy as it occurs, an energy input is required to perform the work of moving the sugar molecules from low to high concentration, much like work was needed to put a ball up in the air.

In a bacterial cell, protons (H^+ ions), as opposed to sugar molecules, are pushed out through the plasma membrane. This means that the concentration of H^+ is greater outside the cell than inside. In the process, the membrane is energized and becomes a storage depot. The membrane is energized because these protons wish to diffuse into the cell just as a ball thrown in the air acquires potential energy and the ability to do work when it descends. Essentially the same process is performed in eucaryotic cells. Protons are pushed out of an organelle called the **mitochondria**. Thus, work can be performed when a proton moves from the cell's cytoplasm into the mitochondria.

Energetic linkage between oxidative reactions and energy storage in membrane.   The advantage of using protons will be explained after we first look at the overall reactions and acquire a sense of the energetics. The energy released from oxidative reactions is used to push protons from low to high concentration. To illustrate further, consider the overall reaction of oxidizing metabolite A to B. When the oxidation of A occurs, energy is released, which can manifest as heat, as in a combustion process:

$$A + \text{oxygen} \rightarrow B + \text{Carbon Dioxide} + \text{Water} + \text{Energy-1 (Heat)}. \quad (4.8)$$

However, when the energy released is used to perform the work of pushing protons up a concentration "hill," the overall reactions and the net result are

$A + \text{oxygen} \rightarrow B + \text{Carbon Dioxide} + \text{Water} + \text{Energy-1}.$
$\underline{\text{Energy-2} + H^+ \text{(low concentration)} \rightarrow H^+ \text{(high concentration)}}$
Net: $A + \text{oxygen} + H^+ \text{(low concentration)} \rightarrow B + \text{Carbon Dioxide} + \text{Water} + H^+ \text{(high concentration)} + \text{(Energy-1-Energy-2)}.$

$$(4.9)$$

If the energy released by the combustion (Energy-1) is completely used for work (Energy-1 = Energy-2), then there will be no net heat release. In practice, some heat release will occur (Energy-1 − Energy-2 > 0) due to imperfect coupling and other effects. However, using the energy released by oxidative reactions to perform the work of pushing protons from high to low concentration lowers the heat release, as the First Law demands.

With H^+ ions at a higher concentration on the outside of the cell membrane, we now have a situation akin to the water stored behind a dam. When a gate is opened and the backed-up water in the reservoir (potential energy source) is directed through a turbine, the potential energy released can be used to perform the work of spinning the rotor in a generator. Likewise, when the H^+ ions on the outside of the cell membrane are allowed to reenter the cell, the energy released with their movement can be used to drive the phosphorylation of ADP to ATP, which requires 7 kcal/mol.

In cells, a special enzyme complex embedded in the cell membrane functions as the "spillway and turbine." It is called the **ATPase complex**. In bacteria, this enzyme resides in the cytoplasmic membrane. In eucaryotic cells, the enzyme is found in the membranes of mitochondria. Recall that for bacteria, the H^+ concentration is greater outside the cell than inside. In mammalian cells, H^+ concentration is greater in the cytoplasm than inside the mitochondria.

Why use H^+ versus other ions or molecules?   A remaining loose end is why H^+ ions are used when in theory, energy can be stored in a gradient of any solute, such as the sugar molecules we initially considered. One reason is that when an oxidative reaction occurs, electrons are removed from the reactant. However, electrons do not "float" around the cell and oxygen concentration tends to be low to prevent oxidative damage to cellular machinery and molecules. Rather, electrons *and* hydrogen are removed from biomolecules as a **hydride** (H^-). The channeling of electrons from reactant to the ultimate recipient, oxygen, is performed by "middlemen" molecules. The first middleman molecule, <u>n</u>icotinamide <u>a</u>denine <u>d</u>inucleotide (NAD⁺; structure shown in Figure 4.2), accepts the hydride from the reactant and in the process, becomes reduced (now abbreviated as NADH). The NADH middleman

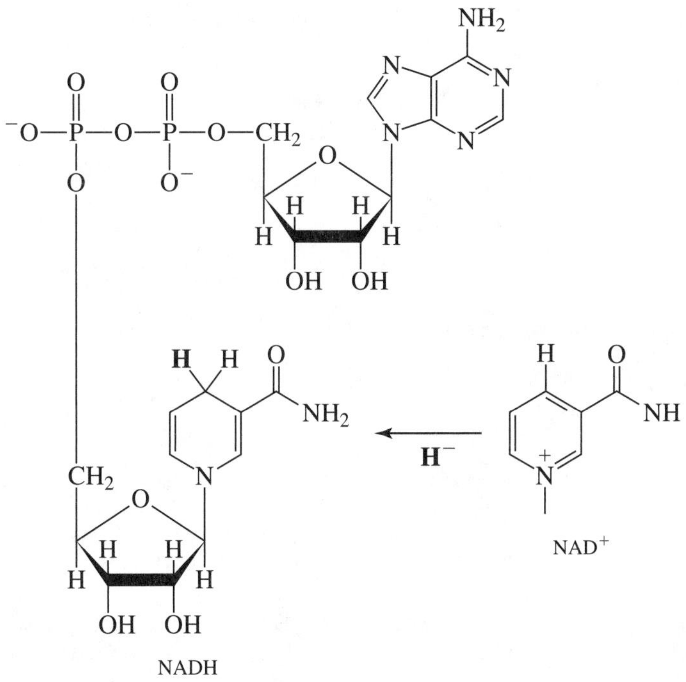

FIGURE 4.2 Structures of oxidized (NAD^+) and reduced (NADH) <u>n</u>icotinamide <u>a</u>denine <u>d</u>inucleotide. The oxidized form accepts a hydride from the reactant in an enzyme-catalyzed reaction. The reduced form conveys the electrons to a membrane-bound charge carrier.

hands off the electrons to the subsequent middlemen and a free H^+ ion remains, as shown in Figure 4.3 for the case of ethanol oxidation.

The middlemen that transact with nicotinamide adenine dinucleotide are proteins imbedded in the membrane. They can directly accept or transfer electrons via the oxidation or reduction of bound metal ions. Thus, "wires" exist in the cell membrane, and nicotinamide adenine dinucleotide initially conveys the electrons to the wires. Each time a transfer of electrons occurs towards oxygen, enough energy is involved to perform the work of "pumping out" H^+.

When taken together, oxidative reactions yield H^+, and some H^+ ions are also pumped out of the cell to provide a difference in H^+ concentration across the membrane. Some of these protons are harnessed by being allowed to re-enter the cell, which can drive ATP synthesis, as also shown in Figure 4.3. The management of the traffic flow of H^+ out of and into a cell by regulatory mechanisms allows for pH and ATP synthesis to be controlled *and* sustained.

There is also an advantage to using a charged solute such as H^+ to energize the membrane. The higher H^+ concentration on the outside of the plasma (bacteria) or mitochondrial membranes, compared to inside, results in an electrical charge difference. "Charging" the cell membrane helps the cell to import a variety of ions, as shown in Figure 4.4. Indeed, cells have many clever uses for the electrical voltage across their membranes, and other charged ions such as potassium and sodium are also used, as you will learn in future biochemistry and physiology courses. Overall, cellular membranes can be viewed as miniature storage batteries that can power tasks such as ATP formation and the importation of ionic nutrients.

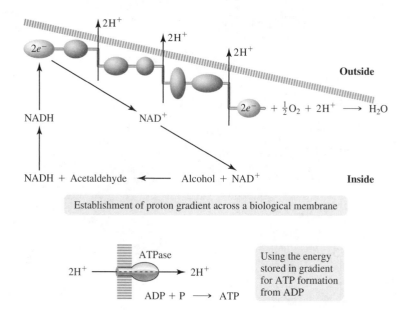

Establishment of proton gradient across a biological membrane

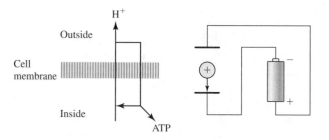

FIGURE 4.3 What occurs when alcohol is oxidized. (Top) Ethanol is oxidized via the enzyme, alcohol dehydrogenase, to acetaldehyde. The electrons from ethanol are carried to oxygen via the NADH and the "middlemen" proteins embedded in the membrane. As the electrons are conveyed in the energetically downhill direction, enough energy is liberated at each downhill step in energy to pump out protons. Overall, the energy obtained from "burning" ethanol is stored in a proton gradient across a membrane. (Bottom) Some of the energy stored in the proton gradient can be used to drive the synthesis of ATP from ADP.

FIGURE 4.4 The effect of pumping out protons is to establish a voltage across the membrane. In addition to ATP synthesis, this voltage can be used to draw into the cell positively charged nutrients akin to the process on the right where a battery establishes an electrical field across two electrodes. This field can cause a positive charge to move.

4.8 Representative Energetic Values at the Cellular Level

Minimal requirements. Earlier, the basal metabolic rate of a human and the magnitude of energy expenditures were benchmarked. We can do the same for cells and compare them to a human. To produce the macromolecules in a bacterium such as *E. coli*, a minimum of 1 mole of ATP is required to build thirty grams of cell material on a water-free basis. This value can be calculated based on the cellular composition and the ATP used in the known biosynthetic pathways that lead to the synthesis of proteins and other major macromolecules. Thus, if a microbe reproduces every hour and there is 7.3 kcal of energy available from a mole of ATP, then a growing bacterium uses a minimum of 240 cal/g h. This estimate is a minimum, because the costs for transporting materials into the cell and other costs have not been included. A human at rest consumes about 1 cal/g h, assuming that the basal rate and mass are 72 kcal/h and 70 kg, respectively.

Efficiency of capturing oxidation energy. From established biochemistry, we can determine that when one mole of glucose is metabolized and oxidized within a cell, about 36 mol of ATP are produced. Assuming each mole of ATP stores 7.3 kcal of energy, there is 263 kcal of energy stored from that originally available in the glucose. Because the complete oxidation of glucose to water and carbon dioxide yields 686 kcal/mol, the efficiency of energy captured by a typical respiring cell is 38 percent. Considering that automobiles and power plants achieve efficiencies of around 25 percent, cells are quite efficient energy capturing systems.

Transmembrane voltage. A rough estimate for the voltage across a mitochondrial membrane is 200 millivolts (mV; 1 mV = 10^{-3} Volts). That may seem pretty weak compared to a 12,000 mV car battery or the 15,000,000 mV peak that exists across the gap of an automobile spark plug when it fires. These comparisons are not fair because the force that a voltage or any potential offers depends on the length scale over which the potential exists.

To illustrate, old horror movies scenes are typically dramatized with the sparking across two electrodes in the laboratory of a mad scientist. If the electrodes were moved apart, eventually the voltage available would not be sufficient to move charge between the electrodes. A related example is the spark plug in an automobile engine. Adjusting the gap is important because if it is too wide, no spark will occur and the cylinder bearing the over-gapped spark plug will not "fire."

That a force formally depends on the distance over which a voltage or any potential is applied can be illustrated with an example from your physics class. Lifting a brick against a gravitational field endows the brick with potential energy given by $U = mgy$, where U, m, g, and y equal the potential that exists, the brick's mass, gravitational constant, and the distance it was raised, respectively. The derivative of U with respect to y provides a measure of how much potential exists per unit of distance, which is also equal to a force. The derivative dU/dy yields mg, the gravitational force that can act on the brick. By analogy, the derivative of a voltage applied across space, dV/dx, provides a meaningful measure of what (electrical) force is available to move a charged entity and also a means to compare two different physical situations. The derivative of a potential is often called a **gradient**.

When the derivative calculation is applied to a mitochondrial membrane, which has a thickness on the order of 100 Angstroms (Å; 1 Å = 10^{-8} cm), we find that there is a "force" of 200 kV/cm available. For a spark plug gap with a typical gap of 0.035 inches, the peak voltage gradient is 1000 kV/cm. The spark plug's gradient is an order of magnitude higher than the mitochondrial value, but the potential carried by a cellular membrane is still capable of performing some significant work, which fortunately does not include sparking.

EXERCISES

4.1   A swimming pool contains 20 m * 10 m * 2 m of water. Estimate how many people have to stay in the pool for one hour to raise the water temperature by 1 degree Centigrade if they are not exercising. If an average person has a volume of 0.72 m^3, is the answer feasible?

4.2   How much water has to be evaporated per hour to dissipate the basal metabolic heat release rate of 72 kcal/h?

4.3   A 70 kg jogger starts from rest and in 3 seconds a pace of 5 km/h is set. The drag force exerted by the air, which a runner must overcome, is given by

$$\text{Drag Force} = F_D = C_D A\ 1/2\rho v^2,$$

where A, C_D, ρ, and v are the runner's frontal area, drag coefficient, fluid (air) density (29 g/22.4 lit), and speed, respectively. Assume that a runner's area is roughly 0.9 m^2 and, in air at the running speed, $C_D \approx 1.0$. Also recall that 1 Newton = 1 kg m s^{-2}; 1 Joule = 1 kg m^2 s^{-2}; 1 cal = 4.2 Joules.

(a) If the runner is able to convert 50 percent of available metabolic energy to work, estimate the kcal/h of metabolic energy used to just overcome air resistance.

(b) Assuming the same efficiency, estimate the kcal/h of metabolic energy needed to get up to the running speed of 5 km/h?

(c) Over a time of 1 hour, how much extra heat has to be removed from the runner's body based on the above estimates and assuming a basal rate of 72 kcal/h? What percentage of the basal rate is released?

(d) Based on your experience with running or observing runners, does this answer make sense? If not, what may be missing in the energy accounting?

(e) The drag coefficient depends, in part, on speed and the density of air. Assume that at typical running speeds, the drag coefficient curve is flat with respect to speed; hence, C_D is constant. So what advice would you give the runner on how to lower the metabolic drain when the running speed is much higher than 5 km/h?

4.4   When 1 mole of ATP is hydrolyzed to ADP, enough energy is liberated to drive the synthesis of 30 g of cellular material, which includes protein, DNA, etc. ATP, in turn, ATP is obtained from the metabolism of molecules such as glucose. One mole of glucose (MW = 180) can yield about 36 mol ATP through substrate- and oxidative-level processes. A can of Coke contains 39 g of sugar. Assuming the sugar in Coke is glucose, how much energy in terms of ATP is in a can of Coke? What total mass (i.e., including water) of cells could be produced based on the energy content of a can of Coke? (Note: Assume the sugar in Coke provides only energy; the carbon to build the mass comes from somewhere else. So consider a can of Coke in terms of energy equivalents).

4.5   In a mechanical system, the potential energy available to do work is $\Delta U = mg\Delta h$, where Δh is the distance a mass is raised above a zero potential energy datum. In a chemical system that has two different solute concentrations, the (Gibbs free) energy that is available to do (chemical) work is

$$\Delta G = RT\ ln\ [C_1/C_2],$$

where R and T are the gas constant (2 cal/mol K) and temperature (Kelvin). C_1 and C_2 refer to the concentrations (e.g., molarities, M) of a solute on different sides of a membrane.

(a) For a one unit difference in pH across a membrane and at body temperature (37°C), what is the energy (in kcal/mol) that is available to do (chemical) work?

(b) This gradient is to be used to drive the reaction $ADP + P \rightarrow ATP$. A concentration gradient of any solute has potential energy. When the solute is charged, a voltage is also established across the membrane, which also adds to the total potential energy. What fraction of the energy needed to drive the reaction is provided by the voltage across the membrane?

4.6   Some drugs can effectively produce "holes" in microbial membranes that permit H$^+$ to leak in and out. Why are such drugs potent antibiotics in that their administration slows microbial growth and thus the course of infection?

4.7 An interesting discovery was made by the University of Massachusetts researchers Daniel R. Bond, Dawn E. Holmes, Leonard M. Tender, and Derek R. Lovley, which was reported in the journal *Science* (Jan. 18, 2002, vol. 295). They placed a graphite electrode in the oxygen-deficient sediment of Boston Harbor. A wire connected the deepwater electrode to another electrode near the surface in oxygen-rich water. They observed that a current passed through the wire with enough magnitude to power a calculator or similar electronic device. Notably, current flow through the wire occurred when microbes were present that could metabolize the organic matter in the sediment. Natural resource and bioengineers are quite excited about this finding because the prospect now exists for building biological batteries that can power undersea observatories and other applications. Use a diagram and words to explain how this battery apparently works.

4.8. A spring stores potential energy (U) equal to $U = \frac{1}{2} kx^2$, where k is the spring constant and x refers to the displacement from equilibrium. For a fixed displacement, how does doubling k alter the force available to do work?

4.9 Which system is capable of exerting a great force: (a) 10 volts established across a 1-mm air gap or (b) 1,000 volts established across a 1-m air gap, and why?

4.10 A muscle cell is oxidizing a compound for energy. The oxygen in the tissue is depleted. Briefly state what will happen to the NADH concentration within the cells once the oxygen concentration drops to zero, and describe why.

4.11 You just ate a 250 Calorie Snickers bar. If it takes about 3 moles of ATP to produce 100 g of protein, roughly how much protein could be theoretically synthesized using the energy content of the candy bar to drive nonspontaneous biosynthesis?

(a) 1 g   (b) 1 kg   (c) 3 g   (d) 3 kg   (e) 5 g

Biomolecular and Cellular Fundamentals and Engineering Applications

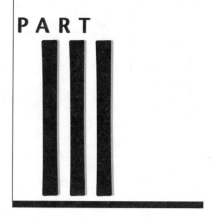

PART

III

Molecular Basis of Catalysis and Regulation

Simple don't mean easy

Sun Record's artist commenting on Johnny Cash's often underappreciated lead guitarist, Luther Monroe Perkins. He used few strings, but got an amazing number of solos just right.

5.1 | Purpose of This Chapter

In Chapter 3, system features and integration requirements were described that included using directed energy investments and feedback control mechanisms. Chapter 4 provided more details on energy transducing mechanisms and how to calculate energetic requirements and expenditures. In this chapter, we will construct a picture of how rate acceleration and coordination are accomplished at the *molecular level* within cells. You will see that even though one process is used, **binding**, there are a multitude of variations that provide flexibility and endow different functions. This simplicity, but with varied outcomes extends across the biological world from bacteria to humans. From the bioengineering standpoint, understanding the mechanisms and outcomes is important to many efforts such as designing drugs, delivering drugs to particular locations in the body, and selecting or developing the materials for artificial organs. The next chapter will deal with the mathematical analysis of binding phenomena.

5.2 | Binding in the Biological Context

It is important to first recall the spectrum of intramolecular interactions. In prior chemistry courses, you encountered covalent bonds, which are the strong links formed between atoms to construct a molecule. Alternately, ionic bonds can form, as in the salt, sodium chloride. Such bonds are important in biological systems and indeed, they are responsible for holding proteins and other biomolecules together.

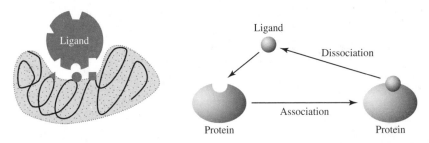

FIGURE 5.1   Binding site on a protein's surface and the reversible binding process. (Left) The folding of the polypeptide chain typically creates a crevice or cavity on the protein surface. This crevice contains a set of amino acid side chains exposed in such a way that they can form strong (noncovalent) bonds only with certain ligands. (Right) The association and disassociation of a ligand with a protein.

When **binding** is referred to in a biological context, the event that occurs is the formation of a noncovalent *and* specific physical linkage between two molecules or between a molecule and an atom. Figure 5.1 provides an example of what is meant by a **binding site** and binding. The large molecule can be a protein, and a localized region (binding site) is shown where another molecule or atom can attach. The attachment is not via covalent bonds; rather, positive and negative charge interactions and other forces hold the pair together.

Because a localized region of the protein can most avidly bind the other molecule, there is **specificity** to the binding process. That is, the surface of the protein is organized such that in one region, the forces that can hold a particular pair together are maximized. Typically, one surface region "recognizes" one specific molecule or atom. For example, some proteins will have a region that binds ATP. Other proteins may have binding sites for atomic ions such as zinc (Zn^{2+}). There are also proteins with multiple binding sites; they can bind two or more molecules or atoms.

When the species that binds to a protein is an atom or small molecule, it is often called a **ligand**. The species that binds to an enzyme, and in the process becomes transformed to another molecule, is called a **substrate**. An enzyme is often named for the substrate and the reaction it transforms. For example, the sugar, glucose, can be modified to another sugar, fructose. In this reaction, the number and type of carbon, hydrogen, and oxygen atoms are the same in the glucose and fructose molecules. However, the way the atoms are connected differs; hence, fructose and glucose are **isomers**. Accordingly, the enzyme that catalyzes the reaction is named, **glucose isomerase**. Here, glucose is the substrate, the reaction is an isomerization, and recall from Chapter 1 that most enzyme names end with the suffix "ase."

5.3 | Binding Is Dynamic

When a protein molecule binds a ligand, that pair is not destined to be permanently attached. The ligand can dissociate, resulting in the protein's binding site being unoccupied. The ligand is then free to attach to another protein molecule. The dynamic equilibrium is akin to the behavior of a weak acid such as acetic acid. When added to water, some H^+ ions dissociate from the acetic acid, but there is also continual association of the free H^+ ions with the anion, as represented in the following reaction (also shown is the analogous case for protein and ligand where *PL, P,* and *L* refer to a protein-ligand pair, a protein without attached ligand, and the unbound ligand molecules, respectively):

$$CH_3COOH \longleftrightarrow CH_3COO^- + H^+.$$

$$PL \longleftrightarrow P + L.$$

For the acid, the greater the tendency for H^+ not to dissociate, the higher the solution pH. Likewise, the stronger the interaction between the protein and ligand, the greater the tendency for the ligand-protein pair to form and remain intact, as opposed to dissociating.

The dynamic character of protein-ligand binding is important in cellular regulation and reaction catalysis. There are some proteins whose catalytic or other function is either enhanced or decreased when a ligand binds. Thus, for a given inventory of proteins, the actual aggregate "work" the proteins perform is based on continually sampling the interior state of the cell and adjusting the output accordingly. The reaction time associated with binding is short; hence, adjusting the protein inventory is deferred to situations when more drastic responses to the environment are needed.

5.4 | Different Venues in Which Binding Operates

Enzymatic catalysis.   For a substrate to be transformed, such as the conversion of glucose to fructose, the substrate must first encounter and attach to the enzyme. A reversible reaction scheme can be written as follows, where *S, E, ES,* and *P* refer to the substrate (e.g., glucose), enzyme molecule (e.g., glucose isomerase), enzyme-substrate complex, and product (e.g., fructose), respectively*:

$$S + E \longleftrightarrow ES.$$
$$ES \longleftrightarrow P.$$

The first reaction represents the binding process whereas the second reaction depicts the chemical reaction (breaking and rearrangement of covalent bonds) that occurs after binding has occurred.

A pictorial view is shown in Figure 5.2. Here, the idea is to depict that a specific region on the enzyme's surface provides a site for binding and reaction. The actual protein structure is more complex than that shown in the cartoon, and many sources exist for displaying computer-generated three-dimensional pictures of enzymes and enzyme-substrate complexes based on X-ray crystallography or proton nuclear magnetic resonance data. Links to some sources are provided on the website.

A question you may now have is "What does binding have to do with catalyzed chemical reactions?" as the second step in the above mechanism indicates. Recall from chemistry that energy is often required to break covalent bonds so that molecular

---

*Here, *P* refers to the reaction product, whereas when ligand binding without reaction was discussed, *P* referred to ligand-free protein. Thus, it is possible to become confused with nomenclature. However, the "P" symbol and dual meaning has become entrenched in many texts. When one sees the symbol "P," the problem context determines if it means reaction product (catalysis) or bare protein (no catalysis).

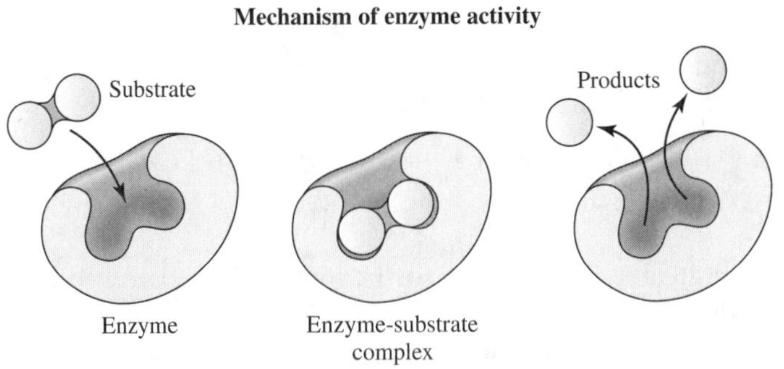

FIGURE 5.2 Schematic of the steps that occur during an enzyme-catalyzed reaction. First, the substrate binds (left) to form an enzyme-substrate complex (center). Reaction occurs, which in this case entails the cleavage of the substrate molecule and the release of two fragments (right).

Mechanism of enzyme activity

Substrate

Products

Enzyme

Enzyme-substrate
complex

rearrangement or other modifications can occur. Not all molecules are in the same energy state, just as not all students in a classroom are equally awake during a lecture. Some molecules have bonds that are more strained near the breaking point than others.

These differences between molecules lead to some molecules being further along in the process of becoming different molecules. A molecule whose structure lies somewhere between the reactant and product state is said to be in a **transition state**. A catalyst can be viewed as an agent that, through its interactions with the reactant, causes some bonds to be stretched or strained so that, they appear to be partially transformed towards becoming a different molecular species. The binding site on an enzyme thus causes distortions in the substrate such that the temperature at which the reaction rate proceeds at an appreciable pace is lowered. A catalyst can also be viewed as an agent that structurally locks in and thus stabilizes molecules in the transition state, thereby guiding them through the passage between reactant and product.

Both views are often expressed by those who study reaction mechanisms, and evidence that supports the second view has recently been presented for the action of the enzyme, orotidine 5'-monophosphate decarboxylase. This enzyme catalyzes the last step in the synthesis of urididine monophosphate, which is one of the four bases found in the DNA. As the name implies, the enzyme removes a carbon dioxide ('carboxy') entity from the reactant, orotidine 5'-monophosphate.

Metabolic regulation and binding.   Binding not only is the prelude to an enzyme-catalyzed reaction, it can regulate the rate of reaction via **feedback inhibition**. A common example is that the end-product of a series of enzymatic reactions can reduce the production rate when the use of the product decreases. A feedback diagram can illustrate the idea, where four enzyme-catalyzed, metabolic reactions occur that sequentially convert $S1$ to $S4$, where S denotes a substrate entity. Such a regulated sequence is often referred to as a **metabolic network**:

The arrows depict chemical reactions, and thus the flow of mass. The dotted line, in contrast, denotes the transmission of information. Here, the concentration of $S4$ is transmitted and "sensed" by the first enzyme in the sequence, akin to how ATP level can influence how fast glucose is used (recall Figure 3.4). The negative feedback aspect is indicated in the diagram with the minus sign in parentheses (−).

The logic behind such an information flow is as follows: When the consumption of S4 declines, it can accumulate much like the water in a reservoir rises when less water is withdrawn through a dam's floodgate. Rather than producing excess S4, the system is designed such that an elevated level of S4 signals the first enzyme in the sequence to slow down the transformation of S1. Consequently, the supply of S4 is kept in balance with the demand.

How this feedback control can occur at the molecular level is shown in Figure 5.3, again in cartoon form. In addition to a binding site on the first enzyme that recognizes S1, another site exists that can bind S4. When S4 binds, the structure of the first enzyme is altered such that the tendency for S1 to bind is reduced. Because the binding of S1 is the prelude to ultimately producing S4 via the sequence, the molecular mechanism that provides negative feedback targets the first committing step in the process.

A lot of terminology is used to describe these binding-based, feedback systems and their variations. One common descriptor is **allosteric enzyme**, which is derived, in part, from the French language and coined by the famous Frenchmen and Nobel Prize recipients, **Jacques Monod, François Jacob,** and **André Lwoff** (http://www .nobel.se/medicine/laureates/1965/). *Allo* means "other" and *steric* suggests "site" or "region of influence." Thus, an allosteric enzyme is one that possesses different binding sites: one recognizes the substrate and the other site(s) binds molecules that can influence the rate at which the enzyme transforms its substrate. Because a protein is an organized yet flexible structure, the different binding events each have a vote on net catalytic rate. The variations are also interesting. Sometimes an allosteric enzyme's catalytic activity is *increased* when a ligand binds, as opposed to the case shown in Figure 5.3 (bottom) and discussed previously. This may occur when a

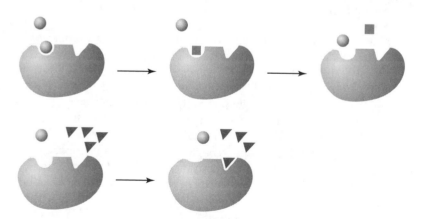

FIGURE 5.3   Representation of how a feedback-inhibited enzyme with multiple binding sites works in a metabolic network. (Top) The protein has one binding site that recognizes the substrate (small circle). Reaction proceeds from enzyme-substrate complex to product (box) formation and product dissociation. (Bottom) When the product from downstream reactions (blue triangle) is very abundant, it has a greater likelihood of binding to the enzyme than the substrate. Once bound, the structure of the enzyme can change such that the enzyme tends to bind the substrate less avidly. Thus, downstream information is used to alter the enzyme rate.

downstream metabolite is in short supply and consequently, an upstream supply source has to be stimulated. For example, if ATP is low, then ADP is elevated. ADP binding to some enzymes will stimulate the ATP-producing reactions.

To round out this discussion, an enzyme need not be of the allosteric type to be susceptible to feedback inhibition. Referring again to the metabolic network shown above, the first enzyme in the sequence, which converts S1 to S2, may have only one binding site; this site primarily binds S1. However, if S4 or another molecule resembles S1 enough so that it can bind to the enzyme's binding site, an elevated level of S4 or the other molecule can prevent S1 from binding. This is known as a **competitive inhibition** mechanism.

From a practical standpoint, many drugs are designed (or found out later) to be imposters of substrates and regulatory molecules. That is, the drug's structure is such that binding to a critical enzyme can occur and thereby reduce the reaction rate. The antibiotic penicillin is a classic example. Penicillin binds to the enzyme that catalyzes the last step in the formation of cell walls in some bacteria that can infect humans and other animals. Consequently, weak cell walls are built which, in turn, results in the pathogens being susceptible to leakage or disintegration (often called **lysis**). Overall, there are many variations, but all mechanisms that are a prelude to enzymatic catalysis, or alter the rate, rely on binding.

Binding in gene expression.   So far we described feedback mechanisms that become more prominent as an end-product (e.g., S4) increases in concentration. Recalling the prior chapter, the example in the above section is an illustration of a continuous variable, negative-feedback control process. The ON-OFF control and its advantages and roles were also described in the last chapter, and as you may expect by now, binding is also central to the operation of these mechanisms.

As discussed in Chapter 1, **gene expression** refers to the use of a particular piece of information encoded in a cell's DNA molecule. Because each **gene** encodes the recipe for building a particular protein, the presence or absence of a particular protein provides evidence for the expression or lack of expression of the encoding gene. To insure that only the proteins needed are expressed, a number of binding-based control mechanisms are used by cells.

The salient features of an ON-OFF gene expression mechanism are shown in Figure 5.4. The information encoded by the gene is translated by the enzyme, **RNA polymerase** (RNApol), in a left-to-right direction. **RNA** is an acronym for ribonucleic acid and **polymerase** refers to the enzyme's activity: the synthesis of a polymer (large molecule). RNApol catalyzes the production of an analogous copy of the gene. This process is akin to when you photocopy a chapter from a book in the library. The book is safely stored in the library for current and future use, and the working copy you made has the portion of the book's information you have deemed to be important at this time. The copy of the gene RNApol has helped to produce is called **messenger RNA** (mRNA). Therefore, gene expression first entails making a working copy, mRNA. Thereafter, translating the working copy to a protein occurs. Thinking mechanistically, if there is no mRNA, then there will be no protein. Therefore, one logical place to exercise control over gene expression is at the level of mRNA synthesis.

As shown in Figure 5.4, binding sites upstream from the gene are often used to control whether mRNA is produced or not. A protein may normally bind to that upstream site and because of its size and shape, it can block RNApol's access to the gene. The blocking protein may also have another binding site on it which recognizes

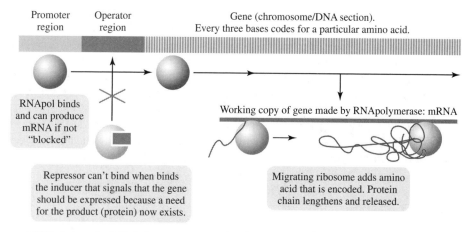

Promoter region

Operator region

Gene (chromosome/DNA section).
Every three bases codes for a particular amino acid.

RNApol binds and can produce mRNA if not "blocked"

Working copy of gene made by RNApolymerase: mRNA

Repressor can't bind when binds the inducer that signals that the gene should be expressed because a need for the product (protein) now exists.

Migrating ribosome adds amino acid that is encoded. Protein chain lengthens and released.

mRNA destroyed quickly to insure gene expression does not occur when switched off by repressor.

FIGURE 5.4   Simplified schematic of how gene expression can be controlled at the level of whether messenger RNA is produced or not.

a ligand that also serves as a signal which indicates that the protein the gene codes for is now needed. When that ligand binds, the structure of the blocking protein is altered such that its ability to make contacts with the gene-flanking site is compromised. Consequently, the tendency for the blocking proteins to dissociate from the gene-flanking site increases with the result that RNApol can access the gene and initiate mRNA synthesis. The blocking protein and the ligand that promotes blocker removal are frequently called the **repressor** and **inducer**, respectively.

There are many variations in how binding is used to regulate gene expression that differ from the scenario in Figure 5.4. For example, in addition to providing access to a gene for RNApol, other binding processes occur that actually increase how avidly RNApol binds to the upstream regions. However, the outcome of all the strategies is similar: proteins that are not needed are not made (OFF) and when they are needed, the binding equilibria shift such that dramatic increase in expression can occur in a short period of time. Therefore, the same basic process, binding, works to manage the protein inventory (gene expression) and how active each protein present is (e.g., end-product binding) at any point in time. Put another way, binding serves to make large-scale adjustments (ON-OFF gene expression) and also underlies the fine-tuning of the system once major adjustments have been made.

Binding in cellular regulation and immune system function.   Binding also plays a role in the transmission of information and the initiation of adaptive responses that involve extracellular molecules. One interesting case corresponds to the use of what may be viewed as reception antennae on the cell surface. Many cells have proteins located on the outer membranes or cell walls. These proteins are capable of binding specific ligands or proteins that are free in the space outside the cell. Such proteins are called **receptors**. One example is the receptors that can bind regulatory molecules such as insulin. Insulin is a metabolism-altering hormone, and some cells produce it while other cells have receptors that bind insulin for the purpose of importing it into the cell for its effect to be realized.

The operation of the immune system also provides an interesting example of binding. A pathogenic bacterium has specific chemical signatures on its cell wall. Some cells in the immune system have receptors that can recognize the presence of molecular signatures that do not belong in the body. When a plasma cell in the blood stream, for example, recognizes a bacterial invader through a binding interaction, that plasma cell is stimulated to divide. That variety of plasma cell and the antibody it secretes then increases. The result is that a large defense against the invader is mounted. The invader is vanquished because the increased number of antibodies coat the invader, which provides a signal for white blood cells to ingest and destroy it. Essentially, the bacterial cells are imported into the white blood cells and dissolved in an acidic organelle.

EXERCISES

5.1. The binding of two different ligands, L1 and L2, to a protein is measured. The protein is found to bind to L1 more strongly than L2. In a solution containing protein, L1, *and* L2, answer the following:

(a) Which will be more prevalent, P-L1 or P-L2, when the total L1 and L2 in the solution are equal?

(b) Will a given protein molecule never bind to L2 after a L1 has bound to it?

(c) Will adding more L2 increase the amount of P-L1 present?

5.2. A metabolic network is shown below. A mutation results in the first enzyme's binding site attaining a shape that fosters increased binding of *S4*. Discuss what the effect on the regulation of the network will be.

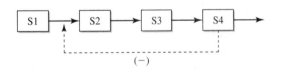

(−)

5.3. In Figure 3.4, ADP is shown to increase the rates of some enzyme-catalyzed reactions while ATP binding can decrease the rates of others.

(a) For reaction 10 in Figure 3.4, draw diagrams akin to those in Figure 5.3 that show how the enzyme-catalyzed rate is regulated through binding.

(b) For reaction 3 in Figure 3.4, assume that the enzyme has two binding sites. One recognizes the substrate, fructose-6-phosphate (F6P). The other can bind either ATP or ADP, but the effect of binding ATP or ADP can differ. With diagrams akin to those in Figure 5.3, show how ATP or ADP binding can alter the rate at which F6P is transformed.

5.4. A mutation has occurred in a cell. One of the repressor protein's binding sites no longer has the right shape to bind an inducer. Discuss what the result will be.

5.5. A mutation has occurred in a cell. One of the repressor protein's binding sites has the right shape to bind to the repressor region regardless of whether an inducer is present or absent. Discuss what the result will be.

5.6. A mutation has occurred in a cell. One of the repressor protein's binding sites has changed such that the shape no longer fosters binding to the repressor region, regardless of whether an inducer is present or absent. Discuss what the result will be.

Analysis of Molecular Binding Phenomena

6.1 | Purpose of This Chapter

The last chapter illustrated that binding underlies a wide range of biological control mechanisms and enzymatic catalysis. Extending this realization leads to the conclusion that diseases or metabolic disorders arise when some binding reactions operate at a "too low" or "too high" level. Drugs or other therapies may then be designed and used to target molecular subsystems that are operating at the "wrong level" in order to restore normal function by either reducing or increasing binding to the "right level." It may also be possible to harness some of the catalysts cells use for technological purposes. For example, the chemical synthesis of a compound in a beaker could be enabled by using an enzyme that has been isolated from a cell. Adding "more" catalyst may speed up a reaction so that it occurs in a "reasonable" length of time.

Exploring these possibilities requires that we can determine what "level," "too low," "more," "reasonable," etc. really mean. To underscore why being more specific is necessary, imagine trying to decide which one of several drug candidates may be the best choice. If knowledge is lacking on how strongly each candidate interacts with a target protein, then there is no basis for estimating what dose should be administered to a patient. Indeed, if the dose is too low, then the target system will be unaffected. If the dose required for an effect is large, then side-effects may occur which may offset the positive effects of the drug. Mechanisms and intriguing possibilities have thus gotten us to the point of needing real numbers to answers tough, interrelated questions.

This chapter covers the basics of how to quantitatively analyze molecular binding in order to answer questions such as "How much of a particular protein binds a ligand?" and "How fast does an enzyme convert a substrate to a product?" Because considerable problem set-up effort is required before useful answers can be obtained, some general comments on how to approach problem formulation and solution are also provided. The utility of the analyses for designing products and processes based on biomolecular binding will be illustrated in the following chapter.

6.2 General Strategy for Problem Formulation and Solution

Students often become frustrated when the instructor shows how to solve a problem and in the process, it appears that some steps are "skipped." What has happened is that after plenty of problems have been solved, a pattern slowly becomes established that provides a general approach to problem solving that can work even when a new problem type is encountered. Because such instincts are slowly acquired through experience, an experienced problem solver's workflow can appear to be arbitrary or opaque to an observer. Thus, we start with some organizational suggestions to use when attacking engineering analysis problems.

There are two basic types of mathematical relationships that are useful to keep in mind when starting and organizing an analysis. **Conservation relationships** are the most rigorous equations that can be written. Conservation relationships are statements that quantities such as mass, energy, momentum, etc. are conserved. These relationships are always obeyed in macroscopic engineering systems, and people tend to not argue about their validity or limitations.

Constitutive relationships are simplified descriptions about behavior that may have a fundamental basis, or simply may be an expression of how something was observed to behave under a particular set of circumstances. An example is how a spring responds when a weight is attached. If you double the weight attached to a spring, the amount the spring typically stretches doubles. You may recall this description of spring behavior as a statement of Hooke's Law. The "Law" is based on experience with many springs and is not based on any details of what the spring is made of; it is simply an encapsulation of reproducible experience.

When very large weights are used, the spring material may undergo stress or alteration such that the simple linear relationship no longer describes the force-extension behavior. Thus, constitutive laws often have bounds over which they work, which is a limitation an engineer needs to be cognizant of. If you base a design on a constitutive relationship that tends to break down under the actual operating conditions, poor performance or worse consequences can result.

Deciding how conservation relationships can be used to link variables is a very good starting point. The constitutive relationships are often used to acquire additional relational information to more completely determine the problem formulation. In problem solving, it may be helpful for you to think and proceed in terms of the following steps:

1. Identify what is known and not known. Identify and define what variables you are specifically seeking to determine.

2. Note what you can say in terms of the applicable conservation, and then constitutive relationships.

3. Note any assumptions that are explicitly or implicitly made by the relationships used. Additional assumptions may be made to simplify the "first-try" analysis. Documenting assumptions at this step will enable you to later evaluate the analysis result and make some refinements, if necessary, in a subsequent analysis.

4. Use the relationships and assumptions to obtain an answer.

5. Subject the answer to limit analyses to see if it makes sense.

6. Evaluate further the answer and use step 3 to remind yourself where some potential weaknesses may lie.

6.3   Analysis of a Single Ligand-Single Binding Site System

Motivation and using the steps.   You probably have had enough of the problem-solving philosophy, so let's take on a real problem by using the above steps. Consider the reversible binding of a ligand to a protein with one binding site. We wish to evaluate what fraction of the protein has a ligand bound as a function of free ligand concentration, given that we know the total protein and ligand concentrations.

This problem has practical dimensions. A good estimate may exist for the total number of a particular type of ligand and protein in a cell. The binding of the ligand has been proposed to activate the protein for participation in a signaling mechanism. Thus, to scrutinize this suggestion, it would be helpful to know what fraction of the protein has bound ligand. If the fraction is small, then the mechanism's validity either is doubtful or the extent of activation is very high. Either way, a calculation will shed some new light on the proposal and help us to formulate the next question to ask.

Let us now follow the previously listed steps:

1. We know the total protein and ligand concentrations. Denote these as (P_T) and (L_T), respectively. We do not know the free protein and ligand concentrations, which we will refer to as (P) and (L). In this problem, the fraction $(PL)/(P_T)$ is sought; hence, define the fraction f as $f = (PL)/(P_T)$.

2. Assume that conservation of mass occurs. The implication is that all the protein has to be in either one of two states: no bound ligand (P), or possessing one bound ligand (PL). A similar statement for ligand can be made.

3. Either a mass or mole balance can be used. We will use a mole balance and indicate in bold what is known.

$$(\boldsymbol{P_T}) = (P) + (PL). \tag{6.1a}$$
$$(\boldsymbol{L_T}) = (L) + (PL). \tag{6.1b}$$

There are now three unknowns $((P), (L),$ and $(PL))$ and two equations, so one more independent equation is needed. This is a good time to consider what constitutive relationships may exist that can link the variables. From thermodynamics, we know that equilibrium constants can be derived and defined that relate what the different species concentrations will be once a binding or reaction process has attained completion. The equilibrium constant K is defined as

$$K = (P)(L)/(PL). \tag{6.2}$$

Note that (P), (L), and (PL) refer to the concentrations of ligand-free protein, ligand that is not linked to protein (i.e., free ligand), and protein-ligand complex*. Moreover, K has been defined to be a dissociation (i.e., PL complex falls apart) constant. This follows from the convention that reaction products appear in the numerator of an equilibrium constant while reactants appear in the denominator. Thus, in Equation (6.2) the products are the "pieces" P and L, while the reactant (starting species) is the complex, PL. Equilibrium constants can be written to represent either association (individual molecules coming together) or dissociation. The use of either association or dissociation constants is mainly a matter of personal choice and can sometimes generate confusion when someone else's result is being used and some decisions have not been noted. The units of K can indicate which perspective was adopted. When the units of K are concentration raised to a positive power (e.g., mol/lit, mol^2/lit^2), then that indicates that a dissociation constant was used in the analysis.

4. Now the problem is completely determined. The fraction we seek can be calculated based on what we know, or a subset of the relationships can be used to calculate f based on what we know and what we are willing to leave as assignable (i.e., L). Noting the definition of what we seek, f, we first eliminate irrelevant variables in terms of relevant or known variables. That is, by using Equation 6.1a, we remove (P) from Equation 6.2 in favor of (PL) and (P_T):

$$K = [(P_T) - (PL)](L)/(PL). \tag{6.3}$$

It is often useful to introduce the target quantity early in the analysis, rather than extracting it from later mathematical manipulation steps. This also focuses your attention on completing the analysis in a way that minimizes the steps and thus the probability of getting lost or making an error. Dividing the numerator and denominator of the right side of Equation 6.3 by (P_T) introduces what is sought, f, and yields

$$K = (1 - f)(L)/f. \tag{6.4}$$

In just one more step, we obtain what we seek, that is, f:

$$f = (L)/[K + (L)]. \tag{6.5}$$

A plot of Equation 6.5 for various values of (L) and K is shown in Figure 6.1. The shape, which begins like a straight line at small values of (L) and then flattens out to an asymptotic limit, is termed **hyperbolic**.

5. To see if the answer makes sense and to flag whether a mistake may have been made, some limits and sensitivity questions can be posed and the answers scrutinized. First, Equation 6.5 indicates that if there is no ligand (i.e., $(L) = 0$), then $f = 0$, which indeed makes sense. Increasing (L) should increase f; we find that as (L) increases and $(L) \gg K$, then $f \approx (L)/(L) \approx 1$; this sensible limit is also obtained. Lastly, what happens as K increases and (L) is fixed, and does it make sense? Recalling the definition of K (Equation 6.2), as K increases, the tendency for free ligand and protein to be present versus

---

*Concentrations of P_1L_1 and PL have been used versus activities, so while K may be rigorously determined from thermodynamics, we will regard Equation 6.2 as a constitutive relationship because the applicability is limited to ideal solutions.

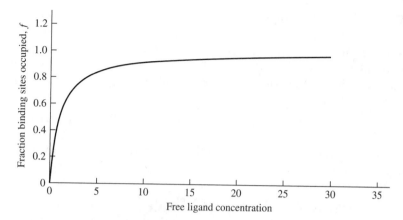

FIGURE 6.1   How the fraction of binding sites occupied depends on increasing, free ligand concentration. In this case, the dissociation constant K was chosen to be 1.0 (arbitrary concentration units).

being complexed together increases. Thus, as K increases for a fixed value of (L), Equation 6.5 predicts that f decreases, which is consistent with our feel for the binding process and the definition of K.

6. In this case, the answer seems to make sense. We can still generate questions on what revisions may be required for this analysis when different situations arise. For example, L in Equation 6.5 refers to the free ligand concentration. How do we set that value in order to calculate f? What would change to analyze binding behavior when the protein binds more than one ligand? What about when the protein is an enzyme which both binds and catalyzes the transformation of the ligand? Such questions will be answered next.

6.4   How to Decide What the Free Ligand Concentration Is

To determine the free ligand concentration, it is necessary to go back to the beginning of the analysis that yielded Equation 6.5. In step 3, there are two balances (Eqs. 6.1a,b). First consider the case when the total ligand concentration vastly exceeds the total protein concentration, $(L_T) \gg (P_T)$. If there was 100-fold more ligand than protein, for example, the balances tell us that when $f = 0$, $(PL) = 0$ so $(L) = (L_T)$, which makes sense. If there is no binding, all the ligand should be free. In the other extreme when $f = 1$, we find from the balances that $(L) = (L_T) - (PL)$ $= (L_T) - f(P_T) = (L_T) - f(L_T)/100 = (L_T)(1 - 0.01) = 0.99(L_T)$. This outcome says that although binding is extensive ($f = 1$), the free ligand concentration essentially equals the total ligand concentration. This outcome also makes sense. The ligand substantially outnumbers the protein; hence, even when complete reaction occurs, there is considerable excess ligand left over as free and unbound. Therefore, when $(L_T)/(P_T)$ equals 10 or more, the error in assuming $(L) \approx (L_T)$ is minor. The error declines as the factor by which the ligand is in excess increases.

When the ligand and protein concentrations are comparable, then the approximation $(L) \approx (L_T)$ is not good, so the bookkeeping should be more thorough in order to obtain an answer with acceptable accuracy. An exact solution can be found by substituting $(L) = (L_T) - f(P_T)$ into Equation 6.5. Here, the exact mass balance equation is used in the analysis, as opposed to a limiting approximation. The result is a quadratic expression for f:

$$f^2(P_T) - f[K + (L_T) + (P_T)] + (L_T) = 0. \tag{6.6}$$

Some examples follow that illustrate binding analysis, and when approximations are acceptable or unacceptable.

6.5 | Examples of Binding Calculations

The total protein and ligand concentrations are 10^{-7} M and 10^{-4} M, respectively. The dissociation constant is 10^{-5} M. What is the concentration of free protein?

> Because $(L_T) \gg (P_T)$, we can assume that $(L) \approx (L_T)$ and use Equation 6.5.
>
> $f = 10^{-4}/(10^{-5} + 10^{-4}) = 0.91$.
>
> Ninety one percent of the protein thus has a ligand bound to it. Because $(PL) = f(P_T)$, the free protein concentration, (P), is $(1 - f)(P_T)$.
>
> $(P) = (1 - 0.91) \, 10^{-7}$ M $= 9 \, 10^{-9}$ M.

If instead the total protein and ligand concentrations are 10^{-4} M and 10^{-4} M, what is the concentration of free protein?

> Because (L_T) *is not* $\gg (P_T)$, we *cannot* assume that $(L) \approx (L_T)$. Thus, the exact solution, Equation 6.6, should be used.
>
> $f^2(P_T) - f[K + (L_T) + (P_T)] + (L_T) = 0$.
>
> $f^2 10^{-4} - f[10^{-5} + 10^{-4} + 10^{-4}] + 10^{-4} = 0$; multiplying both sides by 10^4 considerably simplifies working with the numerical values.
>
> $f^2 - f[2.1] + 1 = 0$.
>
> Using the quadratic formula yields two roots: $f = 0.73$ or 1.37. Only the former is physically sensible because f must be less than or equal to one ($0 \leq f \leq 1$).
>
> The free protein concentration is $(1 - 0.73) \, 10^{-4} = 2.7 \, 10^{-5}$ M.
>
> Note that if the assumption $(L) \approx (L_T)$ were used, f would be found to equal 0.91, which exceeds the exact value (0.73) by 25 percent, which is a significant error. The error in the calculated value of free protein concentration is likewise large.

6.6 Analysis of Binding when Enzyme Catalysis Occurs

Why and using the steps. How much enzyme is needed to convert a substrate to a product in a given amount of time? What will the conversion rate be for a given concentration of enzyme and substrate? These are questions that can be answered by extending the above analysis to include a definition of reaction rate.

Recall from the prior chapter the mechanism of an enzyme-catalyzed reaction:

$$S + E \leftrightarrow ES. \tag{6.7a}$$

$$ES \rightarrow P. \tag{6.7b}$$

The first step, the formation of the enzyme-substrate complex (ES), corresponds to the formation of a protein-ligand complex, which was described above. Different symbols and terminology are used (e.g., *substrate* or *reactant* instead of ligand; E for P; S for L)* to remind ourselves that we are considering a case where binding *and* reaction occur.

We wish to predict the rate at which reactant is converted to product as a function of total enzyme concentration (E_T) and current substrate concentration (S). The conservation and constitutive relationships in this case are

$$(E_T) = (E) + (ES). \tag{6.8a}$$

$$K = (E)(S)/(ES). \tag{6.8b}$$

The rate will be proportional to the concentration of enzyme-substrate complex. This expected dependence is reasonable because (ES) corresponds to the concentration of reactant bound to the active site and thus how much reactant that is prone to molecular rearrangement. Therefore, the rate of reaction (r) is

$$r = k(ES), \tag{6.9}$$

where k is called a **rate constant**. It is worth considering the units of k before moving forward. The quantity (ES) has the units of concentration. Sensible units for rate are how fast the concentration of a species changes with respect to time (e.g., mol/lit-h). The units of k are thus inverse time.

Sometimes k is called a ***turnover number*** or ***turnover frequency***. The idea is that so many substrate molecules (or moles) can be processed per enzyme molecule (or mole of enzyme) per unit time when the reaction rate is enzyme-limited.** A busy restaurant can serve as a physical model of what a turnover number means. Assuming the kitchen is quick and well run, the speed at which the line of waiting diners moves (and rate at which tips and cash flow), depends on the waitstaff's ability to take orders, serve orders, and move the diners out the door so new diners can be seated. A waitstaff person with a fast turnover processes more customers per shift, and also gets more tips.

The analysis should now focus upon obtaining (ES) by eliminating unknown quantities (E) in favor of known $((E_T), K, \text{ and } k)$ and assignable (S) quantities. (ES) can be first eliminated from Equation 6.8b by using Equation 6.8a. With (ES) in

---

*Recall the nomenclature comments and cautions in Chapter 5.

**"Enzyme-limited" means that the number of substrate molecules vastly outnumber the number of enzyme molecules, so the time involved with a substrate finding an enzyme is small compared to the time required for the transformation to occur.

terms of known quantities, Equation 6.9 can be rewritten as Equation (6.10c)

$$K = [(E_T) - (ES)](S)/(ES). \tag{6.10a}$$

$$(ES)/(E_T) = (S)/[(S) + K]. \tag{6.10b}$$

$$r = k(ES) = k(E_T)(S)/[(S) + K]. \tag{6.10c}$$

As performed in the prior analysis of ligand binding, the reader can confirm that a limit analysis (e.g., $S \to 0$) will yield physically sensible results. These details are not repeated here. Rather, it is useful to note the similarities and differences with the ligand binding results in order to fully interpret the sense of Equation 6.10c. Equation 6.10b, which provides the fraction of the total enzyme (protein) that has substrate bound, is essentially that same as Equation 6.5. Because both problems involve the reversible binding of a molecule to a protein, it is reassuring to see that the analyses replicate each other, in part.

Interpreting and generalizing the result.   Equation 6.10c, which expresses the rate of the reaction, can now be viewed as the product of two factors: (1) the rate that would occur if all the enzyme molecules had substrate bound ($k(E_T)$) and (2) the fractional utilization of the total enzyme present ($f = (ES)/(E_T) = (S)/[(S) + K]$). Physically, the first term corresponds to the case when all the binding/catalytic sites are "busy." To use the restaurant analogy, all the tables are occupied and the rate of customer throughput is limited by the waitstaff's table turnover capability. The second term adjusts the maximum rate to the actual circumstances, where the fraction of the enzyme molecules involved with binding or catalysis, based on the relative amounts of enzyme and substrate present, is between zero and one. These two factors are acknowledged in the nomenclature and terms used to describe enzymatic reaction rates. Equation 6.10c is often condensed to

$$r = V_m(S)/[K + (S)]. \tag{6.11}$$

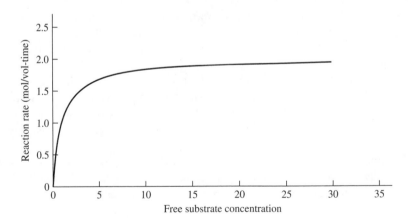

FIGURE 6.2   How the rate of an enzyme-catalyzed reaction depends on free substrate concentration. In this case, K and V_m were chosen to be 1.0 [mol/volume] and 2.0 [mol/volume-time]. Note how an asymptotic value equal to V_m is approached and how the shape of the curve resembles a ligand binding curve.

The parameter V_m is called the **maximal velocity** and has units of rate (e.g., mol/l-s). From Equation 6-10c, V_m is the product of how fast a particular enzyme molecule (k) is and how many enzyme molecules are present per unit volume (E_T). To return to the restaurant analogy, a line of waiting customers will move quickly if there are few, fast waitstaff present or many, slow waitstaff working.

Figure 6.2 shows how the rate of enzyme-catalyzed reaction depends on free substrate concentration (S), K, and V_m. The shape of the ligand binding curve (Figure 6.1) is paralleled by Figure 6.2.

6.7 | A Protein with Multiple Binding Sites

Why and using the steps.   The last case we will consider is when two ligands can bind to *different* sites on a protein molecule. Many cases of multiple ligand binding exist and the ligands can be different or the same. One example was provided in Chapter 5 (Figure 5.3). In addition to a substrate binding site, an enzyme may also possess another site where a ligand (e.g., product of a metabolic pathway) can bind. The ligand binding event, in turn, can either increase or decrease the reaction rate by changing the cooperative interactions in the protein molecule. Recall that this phenomenon of interaction through different binding sites is referred to as **allostery**. A practical example was shown in Figure 3.4. In addition to the substrates binding to the enzymes in the metabolic pathway, ADP or ATP can also bind some enzymes for the purpose of increasing or decreasing a reaction rate.

The oxygen-binding protein in blood, hemoglobin, provides an example of when the multiple ligands are the same. Here, one hemoglobin molecule can bind more than one oxygen molecule, and the tendency to bind or release oxygen depends, in part, on how many oxygen molecules are bound. Again, cooperativity is a factor in that each binding event subtly alters the protein's three-dimensional structure, thereby making it either easier or harder for another ligand to bind.

To illustrate how proteins with multiple binding sites are analyzed, the following example will assume that there are two binding sites per protein molecule. Additionally, the binding constants need not be intrinsically the same, and the state of ligation can affect the tendency for a new ligand to bind. What we seek to compute is how the number of ligands bound (on average) per protein molecule depends on the intrinsic binding constants and the free ligand concentration.

Before beginning an analysis, it may be helpful to diagram the nature of the protein-ligand interaction in order to better appreciate what is meant by "two sites" and "different binding constants." Figure 6.3 depicts a protein molecule with two different sites (I and II) that can bind the same ligand (small circle). A ligand can first bind to either site I or II. Each of the two sites, however, may not have the same affinity for the ligand. After one ligand has bound, another binding event can occur that will occupy the second site. Because of the connections in the protein and the flexible nature of the protein's three-dimensional structure, the avidity of the second binding event may depend on whether site I or II was first occupied. That is, the subtle changes in three-dimensional structure that follow from binding to site I may not be the same as those that ensue when site II is first occupied.

FIGURE 6.3 (Top) Binding of one ligand may affect the conformation of a protein and thus the affinity for another ligand to bind at another site. (Bottom) Some examples of behavior for the binding of a ligand to a protein with two binding sites. When the binding constants are equal, hyperbolic behavior occurs as indicated by the solid curve. When the binding to one site increases the affinity for the binding of the second ligand to increase, a much sharper increase in the number of ligands bound can occur, leading to almost ON-OFF switching behavior (dashed curve).

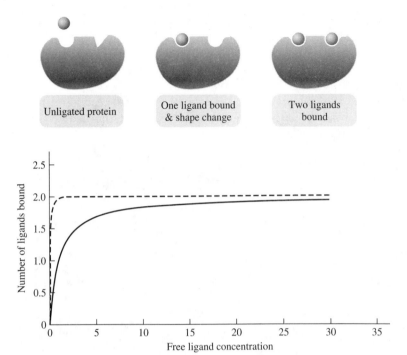

To put the foregoing conceptual model into mathematical form, begin with conservation and constitutive relationships:

$$(P_T) = (P) + (PL) + (PL_2). \tag{6.12a}$$

$$K_1 = (PL)_I/(P)(L). \tag{6.12b}$$

$$K_2 = (PL)_{II}/(P)(L). \tag{6.12c}$$

$$K_3 = (PL_2)/(PL)_I(L). \tag{6.12d}$$

$$K_4 = (PL_2)/(PL)_{II}(L). \tag{6.12e}$$

Unlike the prior use of equilibrium constants (K-values) in Equation 6.8b, the above are defined from the protein-ligand complex association versus dissociation perspective. Again, this is a matter of viewpoint, and association constants are the reciprocal of dissociation constants. Note also that $K_1K_3 = K_2K_4 = (PL_2)/(P)(L)^2$, because the two different paths ultimately lead to the same end point: the binding of two ligands to the protein.

Our target quantity is the number of ligands bound per protein (n):

$$n = \frac{(PL)_I + (PL)_{II} + 2(PL_2)}{(P) + (PL)_I + (PL)_{II} + (PL_2)}. \tag{6.13}$$

Using the definition of n and the relationships for K_1, K_2, K_3, K_4, and K_1K_3 to eliminate (P), $(PL)_I$, $(PL)_{II}$, and (PL_2) in terms of (L) yields

$$n = \frac{[K_1 + K_2](L) + 2K_1K_3(L^2)}{1 + [K_1 + K_2](L) + K_1K_3(L^2)}. \tag{6.14}$$

An examination of Equation 6.14 indicates that sensible limits exist. When $(L) \to 0$, $n \to 0$; that is, when no free ligand exists, it makes sense that the state of ligation of the protein be zero. As (L) becomes very large, $n \to 2$, which is the appropriate upper bound based on the stoichiometry of ligand binding (i.e., there are two binding sites per protein.)

Interpreting and using the result further.   While the limits prescribed by Equation 6.14 may make sense, this equation probably still looks quite complicated and you may be wondering what else it can tell us, or what it is good for. The richness of behaviors endowed by one mechanism, binding, is well captured in the result. The spectrum of behavior depends on the relative values of the association constants. We can thus vary the relative values of the association constants to see what different classes of behavior emerge. In a sense, this is a "pencil and paper" attempt to ascertain what properties that evolutionary processes have conferred to such a system though different molecular designs.

When the binding sites are equal and noninteracting (i.e., $K_1 = K_2 = K_3 = K_4$), we find that Equation 6.14 boils down to a variation of Equation 6.5:

$$n = 2K_1/[\, 1 + K_1(L)].\qquad\qquad(6.15)$$

Instead of an upper limit of 1, the upper limit of Equation 6-15, as (L) becomes large, is 2. Additionally, the equilibrium constant appears in the numerator and denominator in Equation 6.15, whereas it only appeared in the denominator of Equation 6.5. Again, this is because the equilibrium constant referred to association in one problem and dissociation in another. A plot of how n depends on (L) based on Equation 6.15 is shown in Figure 6.3. A characteristic hyperbolic dependence of n on L is exhibited in this case.

More interesting behaviors are exhibited when we allow for unequal association constants, which can arise due to cooperativity effects. Let us now assume that one binding event increases the receptivity for the next binding event. From $K_1K_3 = K_2K_4$, we can say also that $K_1/K_2 = K_4/K_3$. To use some numbers to illustrate how we would assess the effect of such enhanced binding, the second binding event may be 100-fold stronger. Thus, one valid combination of K-values is $K_1/K_2 = 1/1 = K_4/K_3 = 100/100 = 1$.

How n depends on (L) when these relative values are used in Equation 6.14 is also shown in Figure 6.3. The shape of the curve is quite different from the case of $K_1 = K_2 = K_3 = K_4$! The pace of how n increases with respect to (L) is accelerated markedly at low ligand concentration. Indeed, valid combinations of equilibrium constants can be found that provide almost switch-like behavior. Recall from Chapter 2 that one type of control mechanism is ON-OFF, much like the operation of a light switch. This example illustrates how molecular mechanisms in biological systems can lead to almost digital (0,1) or ON-OFF behavior.

Overall, proteins involved in signal and information transmission often have multiple binding sites. Many enzymes also are allosteric, and binding either activates them or reduces their ability to catalyze reactions. An example of one scenario for an enzyme is shown in Figure 6.4.

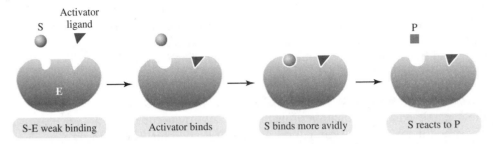

Activator binds and accelerates enzyme-catalyzed reaction

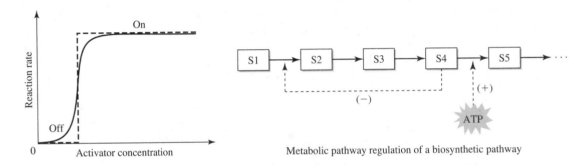

Metabolic pathway regulation of a biosynthetic pathway

FIGURE 6.4   One example of how an enzyme with multiple binding sites can function. (Top) Unless a ligand (triangle) binds and alters the enzyme's structure, the enzyme possesses low ability to bind to S. (Bottom, left) Increasing activator ligand concentration at fixed S concentration will foster more enzyme-substrate binding. Consequently, the rate of the enzymatic reaction can increase in an almost ON-OFF fashion. (Bottom, right) One application of this type of ligand binding can be in a biosynthetic pathway. An enzyme with two binding sites converts *S4* to the biosynthetic precursor, *S5*. ATP binds to this enzyme and increases the avidity of *S4*-enzyme binding. Therefore, unless energy for biosynthesis (ATP) is available, *S4* will accumulate and restrict the rate at which *S4* and *S5* are produced, which is logical because without sufficient energy (ATP) for biosynthesis, *S5* cannot be used. When ATP is available, the conversion of *S4* to *S5* will be stimulated, there will be less negative feedback and increased use of *S5* in biosynthesis.

ADDITIONAL READING

Thermodynamics and Kinetics for the Biological Sciences. Gordon G. Hammes. John Wiley & Sons; 1st ed. ISBN: 0471374911 (June 16, 2000).

Enzyme Catalysis and Regulation. Gordon Hammes. Academic Press. ASIN: 0123219620 (May 1982). (Note: out of print, but if a library copy is obtained, you will find it helpful, succinct, and primarily algebraic.)

Cooperative Phenomena in Biology. George Karreman (Editor). Pergamon Press. ISBN: 0080231861 (March 1978). (Note: dated but shows some prior ideas and analyses.)

Thermodynamic Theory of Site-Specific Binding Processes in Biological Macromolecules. Enrico Di Cera. John Wiley & Sons; 2d ed. ISBN: 0471410772 (July 25, 2001). (Note: may be above the level of some students, but provides molecular and thermodynamic viewpoints.)

EXERCISES

6.1 For the reaction $L + P \leftrightarrow PL$, which of the equilibrium constants below will maximize the concentration of PL when total ligand and protein concentrations are fixed?

(a) $10^{-3} M$ (b) $10^{-4} M$ (c) $10^3 M^{-1}$
(d) $10^4 M^{-1}$ (e) $10^2 M^{-1}$ (f) $10^{-2} M$

6.2 The total protein and ligand concentrations are $10^{-7} M$ and $10^{-4} M$, respectively. The dissociation constant is $10^{-4} M$. Which one of the values below equals the fraction of protein that has ligand bound?

(a) 0.1   (b) 0.2   (c) 0.3   (d) 0.4
(e) 0.5   (f) 0.6   (g) 0.7

If instead the total protein and ligand concentrations are $10^{-7} M$ and $10^{-7} M$, respectively, without resorting to calculations, decide whether the fraction of protein that has ligand bound will be smaller, the same, or greater than above, and explain your reasoning.

6.3 The total ligand and protein concentrations are 10^{-4} and $10^{-6} M$, respectively. What is the free protein concentration when the association constant is $10^4 M^{-1}$?

6.4 The total ligand and protein concentrations are 10^{-6} and $10^{-6} M$, respectively. What is the free protein concentration when the association constant is $10^6 M^{-1}$?

6.5 The association constant for a protein-ligand pair is $10^{-4} M^{-1}$.

(a) When the total ligand and protein concentrations are $10^{-3} M$ and $10^{-5} M$, respectively, what fraction of the protein has ligand bound?
(b) When the total ligand and protein concentrations are $10^{-5} M$ and $10^{-5} M$, respectively, what fraction of the protein has ligand bound?

(c) Redo parts (a) and (b) and let the association constant increase to $10^4 M^{-1}$.

6.6 In a restaurant, 10 waitstaff people deal with 200 customer people in a ten-hour shift. What is the turnover frequency for the "waitstaff catalysts"?

6.7 An enzyme's turnover number and concentration are $5,000 \text{ s}^{-1}$ and $10^{-3} M$, respectively. If the association constant is $10^5 M^{-1}$, what concentration of substrate will yield a rate equal to 2.5 mol $\text{lit}^{-1} \text{ s}^{-1}$.

6.8 A drug molecule can bind to the active site on an enzyme. However, unlike the normal substrate, the enzyme cannot alter the drug molecule. Thus, the drug essentially ties up the enzyme so that less enzyme is available to catalyze the transformation of the usual substrate. The mechanism is

$$E + S \longleftrightarrow ES; \qquad K = [(E)(S)]/(ES)$$
$$E + D \longleftrightarrow ED; \qquad Ki = [(E)(D)]/(ED)$$
$$ES \longleftrightarrow E + P.$$

(a) Show that when so-called competitive inhibition occurs, where S and D refer to the substrate and drug concentrations, respectively, the rate is

$$r = \frac{V_M S}{S + K(1 + D/Ki)}.$$

(b) How would you design experiments and use the information provided by a plot of $1/r$ versus $1/S$ to determine the values of V_M, K, and Ki?

6.9 Describe the effect of the following mutations or drug additions:

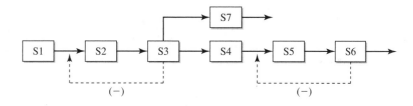

(a) The enzyme that catalyzes the transformation of S1 has a mutated regulatory site such that no binding of S3 occurs.

(b) The enzyme that catalyzes the transformation of S4 has a mutated regulatory site such that no binding of S6 occurs.

(c) A drug that binds to the enzyme that catalyzes the transformation of S2 is added.

6.10. When enzymes are purchased, a data sheet is usually provided. Practical information on storage and database references (e.g. Enzyme Commission numbering) is provided. Nowhere on the data sheet does the turnover number directly appear! Rather "rubber meets the road" numbers are often given instead. These numbers provide users a sense of how much enzyme mass to use in order to get a specific effect. One common number is how many "units" of activity there are per mass of product. The definition of a unit is also typically provided. Based on a molecular weight of 250,000 for catalase and the unit data provided below in the abstracted data sheet, show that the turnover number is on the order of 10^4 seconds^{-1}.

Product Name: Catalase from bovine (cow) liver

EC Number: 2325771

Storage Temp: below 0°C

Unit Definition: One unit will decompose 1.0 μmole of H_2O_2 per min at pH 7.0 at 25°C based on the H_2O_2 concentration falling from 10.3 to 9.2 mM when measured by the rate of decrease of light absorption at 240 nm.

Activity: 2000-5000 units/mg of protein

6.11 If the dissociation constant for catalase-hydrogen peroxide is 10^{-3} M and 5 mg of enzyme is used per liter, answer the following when the turnover number is 10^4 s^{-1}.

(a) What is the maximal velocity, V_M?

(b) Show what the double reciprocal plot ($1/r$ vs. $1/S$) looks like when the substrate concentration (hydrogen peroxide) is varied from 10^{-5} to 10^{-1} M. What does the intercept and slope equal in terms of V_M and K?

(c) What appears to be the utility of such a strange plot when one instead plots $1/r$ versus $1/S$ data for a set mass of newly discovered enzyme that has unknown values of k and K?

6.12 The enzyme catalase breaks down hydrogen peroxide (HOOH) to oxygen and water according to the following reaction:

$$HOOH \rightarrow H_2O + \tfrac{1}{2}\, O_2.$$

For fun, a beaker filled with hydrogen peroxide solution and catalase can be used to inflate an object such as a rubber glove with the oxygen released from the reaction. Assume that the total volume (in liters) of the glove is measured every 30 seconds. When 4 mg of enzyme is used, explain how you would use the volume data and initial hydrogen peroxide concentration to determine the turnover number (seconds^{-1}) and dissociation constant. Recall that the ideal gas law relates gas volume to moles of gas, $PV = nRT$, where P, V, n, R, and T are pressure, volume, moles, a constant, and temperature, respectively. At room temperature (293 K) and atmospheric pressure (1 atm), n (moles) = 4.13 (10^{-2}) V (liters). Your answer should contain both qualitative description (strategy) and "how-to" quantitative analysis (how to process the data).

6.13 Below is a branched metabolic pathway that produces two biomolecules, S4 and S5. The disease, Mellonitis, results from more S4 being produced than S5; hence, the patient has a S5 deficiency relative to S4. Mellonitis results from a mutation in the allosteric enzyme that converts S3 to S5. The mutation results in the binding site having a low association constant for E-S3 formation. Research also shows that the enzyme that converts S3 to S4 is allosteric; binding sites exist for S3 and S4.

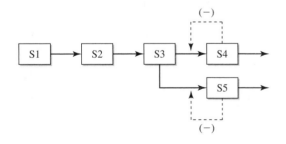

(a) How many enzymes make up the pathway?

(b) How many regulated enzymes make up the pathway?

(c) Will a drug that is a molecular mimic (i.e., analog) of $S4$ possibly reduce the Mellonitis symptoms? Why?

6.14 The first enzyme in a pathway is allosteric and has two sites. Substrate (i.e., reactant S) and ADP can bind to the enzyme. When ADP binds, the enzyme's turnover number increases. Sketch how the enzyme-catalyzed rate depends on S concentration when ADP is (i) absent and (ii) present. Add comments that justify your plot or mark important features directly on the plot.

6.15 It is desired to bind a nonreactive fluorescent ligand to an enzyme. When the ligand binds to an enzyme's active site, the local environment enhances the fluorescence intensity (ligand is brighter). Thus, when the ligand enters the cell, if the enzyme is there and binding occurs, then a fluorescence microscope can be used to visualize cells that possess the enzyme of interest. The enzyme concentration is expected to be around 10^{-8} mol/lit and the association constant is $10^8 \, M^{-1}$. What should the total intracellular concentration of the nonreactive fluorescent ligand be in order for 80 percent of the enzyme molecules to be labeled (i.e., have L bound to them)? State your assumptions.

Applications and Design in Biomolecular Technology

7.1 | Purpose of This Chapter

The background provided in the prior chapters can now be applied. Many technologies have been developed that harness the specific binding that can occur between two molecules. Catalysis may or may not be involved. In this chapter, examples are provided to illustrate the breadth of the uses. Some practical issues that can arise in the design and use of a diagnostic or therapeutic product are also presented.

7.2 | Binding Applications

Fluorescent probe technology for determining type and abundance of molecules on a cell's surface. Recall from Chapter 5 that many cells possess receptors on their surfaces. Molecules present in the extracellular environment bind to these receptors. The binding may serve a communication purpose where, for example, the binding of a hormone triggers a metabolic or gene expression event within the cell. Alternately, binding is a first step in internalizing a growth factor in an animal cell. Because there are many types of cells that can perform different functions, it follows that different cell types will possess different types and numbers of receptors on their surfaces. Thus, one way to distinguish one cell from another is to determine what type of receptors are present on each cell. It can be useful to sort cells when diagnosing a patient, because some diseases lead to different abundances of cell types. Alternately, a normal cell may display different receptor types and/or numbers as compared to a diseased cell. Therefore, detecting the presence of diseased cells at low levels is important for getting a head start on treating the disease before it advances and more complications arise. Many other practical reasons motivate the development of technologies for screening cells for the molecules they have on their surfaces.

Because receptors have specific binding sites, one approach that has been developed to examine cell surfaces for the presence of particular receptors is to attach an easily detectable molecule to the ligand that binds a particular receptor. An effective detection strategy is to attach a molecule that fluoresces to the ligand. Thus, when a modified ligand binds to a receptor on a cell surface, there will be a "glowing spot" located at the receptor-ligand pair. Alternately, a fluorescent-functionalized ligand that permeates a cell can be used to detect the presence and spatial organization of a particular protein. A microscope that can then be used to view the cell population; the cells that glow will only be those that possess the receptor or protein of interest. Figure 7.1 illustrates a labeling strategy and an example of what **fluorescence imaging** technology can yield.

Cell sorting by flow cytometry.   Another application of using fluorescently functionalized ligands is the physical sorting of cells. Rather than looking at cells under a microscope to see what types are present, this technology automates the counting of cell types and the physical separation of one cell type from another. The principle is summarized in Figure 7.2. In one type of sorting application, only one type of cell will bind a fluorescently-labeled ligand whereas the other cell types will not. Thus, only one cell type will emit fluorescence when exposed to a light source (laser) as it flows by. This cell and others like it are detected and flagged for isolation. Isolation is accomplished by using the charged plates to steer the charged cells into one test or another. Because most cells possess a net negative charge, a given cell can be steered to the right or left collection tube by simply by making the left electrode negative or the right electrode negative, respectively. The isolation of primitive stem cells from blood or other tissue sources is accomplished this way. These isolated cells are then used for tissue engineering and other studies.

Drugs that jam enzyme- and protein-mediated processes by binding.   As you get older, you may find that eating a pizza an hour before you retire can be a painful experience. As acid is secreted in the stomach to digest the pizza, some acid may pass up the esophagus. The horizontal position assumed when trying to sleep facilitates the **acid reflux** further. The burning sensation is quite uncomfortable and chronic acid reflux can lead to **erosive esophagitis**, which means that tissue damage occurs. Secreted stomach acids also can aggravate ulcers, thereby slowing or preventing healing.

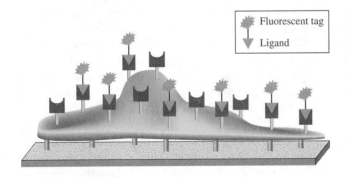 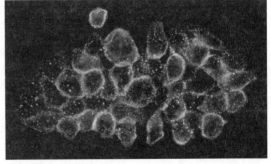

FIGURE 7.1   Using the binding of fluorescently-labeled ligands to establish the types and relative numbers of different receptors present on a cell surface. (Left) A ligand that binds to a particular receptor has been attached to a fluorescent molecule ("tag"). When ligand-cell surface receptor binding occurs, the cell surface acquires fluorescence that can be visualized by fluorescence microscopy or detected by flow cytometry. (Right) Human carcinoma cells exhibit surface fluorescence in a microscope's view field from a probe that interacts with epidermal growth factor receptors. Image G000690 reproduced with the permission of Molecular Probes Inc, Eugene, OR.

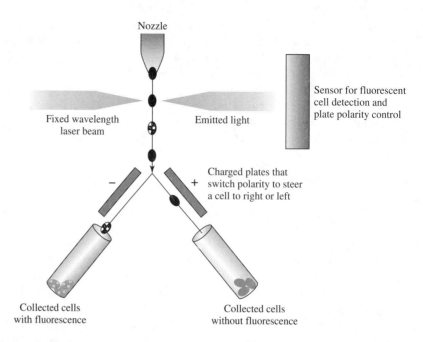

FIGURE 7.2   Schematic of how fluorescence activated cell sorting is performed. A mixed cell population that has been pre-exposed to a fluorescent reporter passes single file through a flow chamber. Light is passed through the chamber that can cause fluorescence to emit from only the cells possessing the distinguishing feature the reporter binds to. Those cells that are sensed to present the fluorescence signature are flagged and physically isolated from the others by using charged plates to push them into the left test tube.

The enzymes that secrete acid into the digestive system resemble the ATPase that was presented in Chapter 4 when membrane energetization was described. Recall that an H^+ gradient can be used to drive ATP synthesis via an ATPase. That is, an ion gradient running downhill releases energy that can be used to drive the reaction, $ADP + P \rightarrow ATP$. The opposite reaction is also possible. ATP hydrolysis can be used to drive H^+ from one compartment into another. Here, H^+ ions pass through the membranes of the **gastric parietal cells** to where the food is located.

The drug **omeprazole**, which has the trade name **Prilosec**, provides an effective treatment. The drug works by inhibiting the H^+/K^+ ATPase enzyme system at the secretory surface of gastric parietal cells. The drug binds to the enzyme and slows down its rate, thereby retarding the rate and extent of acidification of the digestive system. You can learn more about Prilosec and acid reflux by exploring the links provided on the website.

Many drugs work by targeting a specific enzyme and, in essence, altering the enzyme's activity either downward or upward depending on what effect is needed to alleviate symptoms and treat a problem. Antimicrobials (antibiotics) also can function in this manner. The common antibiotic penicillin inhibits the enzymes some bacteria use to synthesize cell walls. The result is that weak cell walls are built and the bacteria disintegrate.

The challenges scientists and engineers face in discovering, developing, and producing such drugs are extensive, but also intellectually stimulating. For example, a dose must be effective, but have low side effects. That means the molecule must be designed so that it seeks and acts solely upon its target. The structures of biomolecules and drug molecules are the "blueprints" that are used to attain specificity. Figuring out how to get the drug to the target is another challenge. If a drug is ingested, it has to pass through the acidic digestive system before entering the blood stream. A drug that looks promising based on chemical and molecular structures may prove to be ineffective in practice because it cannot survive the journey from mouth to target. Injecting a drug into the blood stream is an alternate route of administration. That route, however, is inconvenient and could lead to infection. Thus, drug designers and producers engineer the drug molecule and the components that package it such that (1) the drug's mechanistic effect is preserved to the greatest extent possible, and (2) it can be delivered to its target in an active form. The science of the fate of drugs and other substances in the body is called **pharmacokinetics**, which will be covered in more detail in a later chapter.

7.3 | Enzyme Catalysis Application

The prior examples involved ligand-protein binding. The following technologies use the catalytic capabilities of enzymes.

Fluorescent probes that report on cellular enzymatic reactions.   The spectrum of enzymes a cell possesses is indicative of its metabolic capability and what genes are being expressed. Similarly, the absence of a particular enzymatic activity could be indicative of a mutation, disease, or other alteration in function. Thus, fluorescent probes have also been developed that can indicate if a cell is executing a particular enzyme-catalyzed reaction. Some probes are not fluorescent initially. Rather, they mimic the substrate, or possess a moiety that can bind to a particular enzyme and undergo reaction. After a particular enzyme has transformed the probe, a fluorescent product results. Therefore, the rate and extent that fluorescence appears is indicative of the presence and quantity of a particular enzyme.

One example is provided by the dye **Cyano Tetrazoilum Chloride (CTC)**. If cells are respiring (i.e., producing NADH), then CTC can be reduced by cells such as fibroblasts when an intermediary electron carrier, Medola Blue (MB), is present. It is believed that MB permeates the cell and can then shuttle electrons to CTC, which tends to sequester around the cell membrane. When CTC is reduced, small fluorescent crystals are formed that associate with individual cells. Thus, if a cell glows after being irradiated with light that can excite the crystal fluorescence, it can be concluded that respiratory enzymes are present and the cell is actively engaged in respiration. Interestingly, differences in the fluorescent intensity of cells can be observed when viewed under a microscope that "sees" fluorescence (see Figure 7.3). The intensity differences may relate to variations in the respiratory activity of individual cells. Different cell morphologies may also be responsible for intensity variations. Flat or round cells will protrude into the focus plane differently, and thus individual cells may appear different when viewed under a microscope. Overall, reactive fluorescent probes are a powerful tool for studying cell physiology, the effects and mechanism of drug action, and other applications. Individual cells or small samples of a population can be examined with a microscope, or large populations of cells can be screened via flow cytometry.

FIGURE 7.3 Epifluorescent image (top) and map of fluorescent intensity (bottom) after incubating fibroblasts for 10 minutes with the dye, CTC. The individual cells fluoresce, but with different intensity. Image digitization provides a way to represent view fields as intensity maps. Each peak corresponds to a fluorescent signal. Images acquired by the student, Duc Nguyen, while conducting research in the author's laboratory.

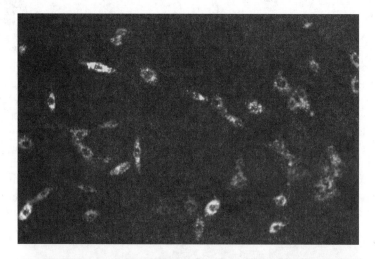

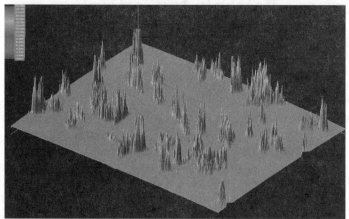

Using enzymes for concentration measurement in diagnosis and disease management.   The control of blood glucose concentration so that it is not too high or low is compromised in people who contend with diabetes. Consequently, regular measurements are made to determine the progression of the disease and to ascertain how well the medication dose and administration schedule are working. A chemical species such as glucose whose concentration is sought is called an **analyte**.

The glucose molecule does not possess a "color" or another property that enables its concentration to be easily measured. Thus, the analyte, in this case, has to be transformed into a more readably detectable substance. One way to measure the glucose concentration in a blood sample is to use the enzyme **glucose oxidase**. The enzyme oxidizes glucose to form hydrogen peroxide. The hydrogen peroxide product can be readily measured by optical or electrical methods. One optical method is to convert the hydrogen peroxide further to a colored product using another reaction. Alternately, the hydrogen peroxide can be reduced to water, and the electrical current drawn to sustain the reaction is indicative of how much glucose was originally present in the blood sample. The reaction can also alter the electrical conductivity of the small test volume, which can be electrically detected.

The enzyme-based strategy used to measure the concentration of glucose in the blood can be applied to other analytes that have medical significance. One needs to find one or more enzymes that can convert the analyte to something that is more easily detected and measured by color-generation or other properties. The development and sale of test kits is a large industry that supports research, veterinary medicine, and human health care.

Engineers who are knowledgeable about biomolecular reactions contribute to the design and improvement of these kits. They strive to exploit the capabilities of enzymes to develop faster and more sensitive measurement strategies. Stabilizing the enzyme components against deactivation, so that the kits can be stored and then used when needed, is another facet of kit design. Finally, considering how someone will use a test kit and instilling the features of simplicity and convenience is another key endeavor. The consideration of human and other factors in design is considered further in the example below.

Design of blood glucose analysis kit: human factors.   Consider developing a kit that will enable diabetics to perform their own measurements of blood glucose concentration. An example is the enzyme reaction-based device manufactured by Bayer http://www.bayercarediabetes.com/prodserv/products/glucelite/, which is shown in Figure 7.4. As an engineer, you have to understand the science as well as make some enlightened design decisions in order for such a kit to be a successful tool for disease management. Some issues in this problem also emerge for other analyte detection kits based on enzyme-catalyzed reactions.

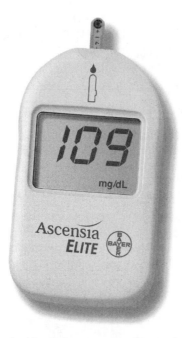

FIGURE 7.4   An example of a blood glucose measurement device. A small blood sample is inserted in the top. The digital display provides the concentration of glucose in the blood sample. The device is about the size of a pager.

One way to begin a design analysis is to think about the human factors and the inherent trade-offs that can arise. Sometimes major issues can be anticipated but often, new ideas emerge as the analysis proceeds. For example, we can anticipate from the *science* (Equation 6.10) that the more enzyme used in each measurement, the faster the reaction will occur and the sooner an answer will be obtained. However, from the *engineering* standpoint, if the enzyme is costly and thus each measurement or the kit is expensive, we wonder if the patient will tend to skip some measurements to save some money. That behavior would raise the risk that they do not monitor their situation well. In the other extreme, using little enzyme in order to save money may prolong the time required to obtain a blood glucose measurement. Thus, during a busy day or when a schedule suddenly changes, the patient may again elect to not make a monitoring measurement. Both design extremes have the same downside: due to lack of regular use, the medication regime used may tend to be haphazard and not result in optimal disease management. Of course, these trade-offs depend on how much the enzyme impacts the total costs, versus the blood sample collection system, the packaging, and so on.

Design of blood glucose analysis kit: engineering analysis.   The downsides of a poorly conceived design are serious. Thus, it is smart to base a design on an improved understanding of the magnitudes of the trade-offs. Assuming that the amount of enzyme used and/or the measurement time are key considerations, we can codify these considerations by defining the quantity ϕ to be the product of the total enzyme used per test kit (E_T) and the time required for 95 percent of the glucose to be reacted to a detectable product (τ). A large value of ϕ is not good; a large mass of enzyme is used and/or the time to wait for an answer is long. Therefore, ϕ provides a reasonable metric for comparing different designs, and it is actually desirable to develop a design that minimizes ϕ. To proceed further, the design problem can be broken down into two parts: (1) determine how ϕ depends on specific enzyme properties, and (2) determine how to make choices such that ϕ is minimized.

We can find the time part of ϕ, τ, by using Equation (6.9c) as a starting point:

$$d(S)/dt = -k(E_T)(S)/[(S) + K]. \tag{7.1}$$

Equation 7.1 states that the rate at which the concentration of substrate (S denotes glucose in this case) decreases is equal to the rate of the enzyme-catalyzed reaction that transforms the substrate to a (detectable) product. To find τ, it is useful to recall some basic ideas learned in physics. Velocity (v) is the first derivative of position (x) with respect to time (t); $v = dx/dt$. When velocity is known as a function of position, then the time needed to cover a given displacement is

$$\int dt = \int dx/v(x). \tag{7.2}$$

This strategy of grouping similar variables on one side of an equation is known as **separation of variables,** which was also introduced in Chapter 2. You will learn other problem-solving techniques in future calculus courses (e.g., Differential Equations). To reduce the "busyness" of the equations, (S) and (E_T) can be "slimmed down" to S and E_T. By analogy to Equation (7.2), the variables can be separated as follows:

$$\int dt = 1/kE_T[K \int dS/S + \int dS]. \tag{7.3a}$$

$$\text{at } t = 0, S = S_o. \tag{7.3b}$$

$$\text{at } t = \tau, \ S = 0.05 \, S_o. \tag{7.3c}$$

The solution subject to the initial value and time point of interest expressed in Equations 7.3b and c is

$$\tau = 1/kE_T [K \ln S_o/S + (S_o - S)]. \qquad (7.4a)$$

To achieve 95 percent conversion of S (i.e., $S = 0.05\ S_o$ at $t = \tau$), the required time is

$$\tau = 1/kE_T [K \ln 20 + 0.95\ S_o]. \qquad (7.4b)$$

Now using the definition of ϕ, we find that

$$\phi = [K \ln 20 + 0.95S_o]/k. \qquad (7.5)$$

Because we can do nothing about S_o, the strategy to simultaneously minimize the product of assay time and amount of enzyme needed is to choose an enzyme that has a large value of k. Because enzymes from different biological sources can catalyze the same reaction yet exhibit different kinetic properties, the search for the "best" enzyme often takes test kit designers into biochemistry databases.

Considering enzyme stability in design.   A first pass analysis often needs to be reevaluated to insure that other important factors are also taken into account. In the design of products made from biological molecules, whether the products are enzyme-based measuring systems or therapeutics, a major concern is **stability**. Here, enzyme stability refers to whether the value of kE_T today will be the same value exhibited six months from now, when the kit may actually be used. Over six months, the enzyme molecules may unfold or become chemically modified such that the same mass of enzyme is present, but only a fraction is in an active form (i.e., the effective value of k has decreased). Therefore, if an enzyme has a high turnover number (k) and a low value of dissociation constant (K), but the shelf life is short, then the attributes of this particular enzyme may never be realized in a product that conceivably is not used immediately after being manufactured. So some further reflection suggests that at least one more trade-off needs to be considered in the design of an enzyme reaction-based, analyte measuring device.

Enzyme deactivation can often be described by a first-order process, which you may have encountered in the description of radioactive decay in prior chemistry or physics courses. In this case, the effective turnover number for some quantity of enzyme decreases exponentially:

$$k = k_o \exp(-k_D \tau_S), \qquad (7.6)$$

where k_o, k_D, and τ_S are the initial value of turnover number, deactivation rate constant, and expected time the kit will be on the shelf and/or in service, respectively. Recalling that half-life (τ_H) and deactivation rate constant are related by $\tau_H = \ln 2/k_D$, we can rewrite what the apparent value of the turnover number (k_{app}) will be at the end of some assumed time on the shelf:

$$k_{app} = k_o \exp(-0.693\ \tau_S/\tau_H). \qquad (7.7)$$

The advantage of rewriting Equation 7.6 in the form of Equation 7.7 is that we now have two convenient yardsticks explicitly stated: (1) our assumption about the longest time the test kit will remain unused (τ_S) and (2) an enzyme's half-life (τ_H). Therefore, the objective to minimize ϕ now becomes

$$\phi = [K \ln 20 + 0.95 S_o]/[k_o \exp(-0.693\, \tau_S/\tau_H)]. \tag{7.8}$$

Now we see that for a fixed value of S_o, there is a **nonlinear interaction** between four parameters: k_o, τ_S, τ_H, and K. In general, we would want an enzyme to have a small value of K. However, a full calculation would be needed to sort out the trade-off between turnover number and stability. For example, if the turnover number was extremely high, an unstable enzyme could possibly prove to be more economical than a lower turnover enzyme with greater stability. Examining these trade-offs further will be the subject of homework exercises.

Other factors in design.   The design process thus far has focused on linking use to the characteristics of the enzyme used. That alone is challenging and introduces interesting trade-offs, but there are other human factors that can make a good design concept succeed or fail. To use many blood glucose analyzers, the diabetic must acquire their own blood sample. The withdrawal of even a small volume of blood from a finger puncture is not a pleasant experience for most people. Thus, substantial conceptualization and engineering design effort is also focused on lowering the real and perceived trauma that results when a patient provides a blood sample to the analyzer. Skin puncture methods that provide small samples with little pain have been developed. Other designers instead strive to build devices that are imbedded in the patient so that a blood sample does not have to be obtained each time a measurement is sought. Rather, the vision is that once the device is implanted, the blood glucose concentration is continuously outputted.

7.4 | Using Enzymes in Food Processing

The food industry uses enzymes to process tons of sugar each year. Although many people take sugar for granted, raw and semi-refined sugars are a large component of international trade (e.g., see http://www.sugarinfo.co.uk) and current geopolitics. The sugar industry has a long legacy of being a significant factor in world history and politics. Indeed, sugar was one issue that led to the American Revolution. In 1764, which was twelve years before the famous 21-year-old Connecticut citizen, Nathan Hale (1755–1776), was hung by the British Army, the British Parliament imposed the first of a series of unpopular taxes, the Sugar Act. Thereafter, relations between the enterprising North American colonists and Britain spiraled downward. A year after the Sugar Act was imposed, the house of Massachusetts Governor Thomas Hutchison was burned as a protest to the next attempt to regain influence over growing North American commerce, the Stamp Act (1765).

Plants provide the raw materials for the sugar industry, which are starch and the sugar sucrose. A **sucrose** molecule consists of two simple sugars, **fructose** and **glucose**, connected together. Fructose has the same molecular formula as glucose ($C_6(H_2O)_6$), but the atoms are arranged differently. Fructose and glucose are thus **isomers**. **Starch**, in turn, is a polymer of glucose molecules. Fructose tastes sweeter than sucrose to the human palate.

Converting raw sugar or starch materials to fructose is economically attractive because less total mass has to be added to a food or beverage in order to gain the same level of sweetness. Enzymes are used to catalyze the conversion to fructose. For example, the major commodity **High Fructose Corn Syrup (HFCS)** is produced by first degrading cornstarch to free glucose molecules. Another enzyme is then used to rearrange the liberated glucose to form fructose.

A large industrial enzyme market exists to supply invertase, glucose isomerase, protein degrading enzymes (used in detergents), and other enzymes. The scale of the HFCS industry is likewise immense. *One* new sugar refinery in Shanghai China, which began operation in 2002, has an annual capacity of 100,000 metric tons of high fructose corn syrup (http://www.china.com.cn/english/investment/19737.htm).

7.5 | **Bioresource Engineering**

There are other uses for the starch found in plants. When starch is degraded to sugar, the sugar can be fed to microbes in order to produce **ethanol** via **fermentation** using large bioreactors akin to the one shown in Figure 0.2. Ethanol production is desirable because ethanol can serve as a fuel or fuel supplement for automobile and other engines. More can learned about the production, advantages/disadvantages, and end uses of alternate fuels from the **Alternate Fuels Data Center** (http://www.afdc.nrel.gov/altfuel/ethanol.html).

Another common polymer found in plants is **cellulose**. Cellulose resembles starch in that it is a polymer of sugar molecules. The bonds between the sugar molecules are different than those in starch; hence, not all animals can derive sustenance from cellulose because the bonds cannot be broken. Humans cannot degrade cellulose, whereas cellulose degradation naturally occurs when, for example, a tree dies and then microbes and fungi indigenous to soils degrade the tree. Degradation occurs because unlike humans, some microbes and fungi that dwell in soils possess the enzyme **cellulase**. This enzyme breaks cellulose down into sugars that can be further metabolized. While the activity of cellulase-producing microorganisms enables nutrient recyling in the environment, which is beneficial, another natural path of cellulose degradation can be quite inconvenient to humans. When termites invade the wood framing in a building, microbes within their guts also possess cellulases. Consequently, with the assistance of gut microbes, termites can rapidly degrade the integrity of a dwelling's wood framing. Again taking a cue from nature, bioengineers also develop processes that use cellulases to convert cellulose to sugars.

When the large panorama of natural materials is considered, some engineers envision that plant-derived chemicals can be used to produce many products that are now produced from petroleum. They envision that a sustainable system and economy can be created where plants (more specifically, carbohydrates) provide a renewable source of raw materials. Improving the processes that harness the capabilities of cellulases and other enzymes is integral to ongoing work that is exploring this vision. Engineers who work on converting biomass materials to useful chemicals and materials are known as **Bioresource Engineers** or **Biocommodity Engineers** (to learn more, see: Lynd, L. R., Wyman, C. E., and Gerngross, T. U. (1999)).

7.6 Immobilized Enzymes in Chemical Weapon Defense and Toxic Chemical Destruction

New uses are always being investigated and developed for enzymes because of the specificity they offer and the advantages afforded by speeding up the rate of a chemical reaction. Using enzymes to degrade **nerve agents** that could be used in warfare or terrorism events is a new application now under investigation.

Examples of nerve gases are **soman** and **sarin**. These agents are known as **organophosphorous compounds** because of their chemical structure. As the name suggests, these agents are organic molecules that contain phosphate. Organophosphorous compounds exert their deleterious effect by inhibiting or deactivating the enzyme **acetylcholinesterase**, whose activity is essential for the proper function of the nervous system in animals. Significant exposure to organophosphorous compounds may lead to convulsions, impaired cardiac and respiratory function, and possibly death.

The enzyme **organophosphorous hydrolase** can catalyze the destruction of organophosphorus compounds. The technological challenge is to find a way to use the enzyme so that it can provide decontamination in a practical setting while still remaining active and stable. This is a challenge because enzymes are normally found inside cells or associated with cellular membranes, which provide them with protection from unfolding or other activity-reducing processes while they perform their metabolic tasks. Therefore, using enzymes in a radically different environment may result in no reaction catalysis or only a short-lived burst of capability.

Immobilized enzyme technology attaches enzymes to a material in order to endow the material with a desired catalytic capability. Apart from endowing catalytic activity, another potential advantage is that the enzyme-material combination can be reused, as opposed to using new enzymes every time the reaction is carried out. However, the chemistry and material science need to be managed so that the enzyme is strongly attached to the material while the three-dimensional structure of the enzyme molecule is "pinned down" and thus stabilized. This management presents a

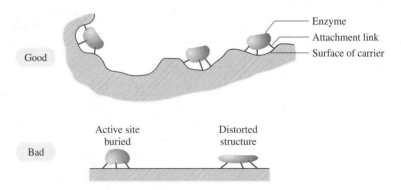

FIGURE 7.5 Illustration of "good" versus "bad" enzyme immobilization outcomes. (Top) In the "good" outcome, enough protein-surface attachments are provided to anchor the enzyme to the surface without overly distorting the enzyme's structure or burying the active site. (Bottom) In the "bad" outcome, the enzyme's structure is distorted and the active site is inaccessible to substrate molecules.

very delicate engineering problem at the molecular level, as Figure 7.5 shows. If an enzyme is excessively attached to a surface by chemical bonds or other means, its structure may be stable, yet so distorted that the catalytic capabilities are markedly reduced.

Organophosphorous hydrolase has been shown to function in specially prepared polyurethane sponges and special clothing. The enzyme molecules are strongly bonded to these materials while still retaining the ability to degrade nerve agent-like molecules after being stored. The choice of sponges and clothing was based on trying to develop new tools that military or civilian response teams can use to safely enter and decontaminate an area that has been exposed to nerve agents.

Some envision that the use of enzymes in nerve agent decontamination can be extended to pollution control. Stores of toxic chemicals exist worldwide. Storage in tanks temporarily prevents the release of these chemicals into the environment. Ultimately the chemicals must be destroyed before an accidental release occurs through tank corrosion or rupture. Thus, enzyme technologists are trying to develop immobilized enzyme processes to convert toxic molecules into less toxic molecules. Indeed, another use of immobilized organophosphorous hydrolase is to reduce the inventory of old or banned nerve agents in military installations. This is an endeavor with high potential impact, because it is estimated that the worldwide stores of organophorphorous compounds exceed 200,000 tons.

REFERENCES

Biocommodity Engineering. Lynd, L.R., Wyman, C.E., and Gerngross, T.U. *Biotechnology Progress*. 1990. *15*(5): 777–793.

Covalent binding of a nerve agent hydrolyzing enzyme within polyurethane foam matrices. LeJeune, K.E. and Russell, A.J. *Biotechnology & Bioengineering*, 51(4): 450–457, 1996.

Degradation of chemical weapons using enzymes in fire fighting foams. LeJeune, K.E., Wild, J.R., and Russell, A.J. *Nature*, 395: 27–28, 1998.

EXERCISES

7.1 A fluorescence-activated cell sorting (FACS) procedure fails to yield cells that bear a particular receptor from a tissue sample. These cells are important to your tissue engineering work, so you need to figure out what the problem is. You suspect that the cells are really there and the fluorescently-labeled probe chosen does not work as well as you expected. Time on a FACS machine is hard to arrange and can be expensive. What is an alternative route you can use to debug this problem and how would you proceed?

7.2 The dissociation constant for a fluorescently-labeled probe is 10^{-6} M. The probe is to be used to detect the presence of stem cells in umbilical cord blood and to isolate them by flow cytometry. The total number of cells in the 5 ml blood sample is 10^8 and it is estimated that 0.1 percent are stem cells. Each stem cell is estimated to possess 100 receptors to which the fluorescent probe will bind.

(a) How much total probe in mol/lit has to be added for fifty percent of the total receptors to have fluorescent probe bound?

(b) Which is more likely, fifty percent of the stems cells will be fluorescent or fifty percent of the receptors on a given stem cell will "glow"?

7.3 The properties of three different fluorescently-labeled ligands are given below.

(a) What are the trade-offs (i.e., relative good and bad features of the different ligands)?

(b) Which would be the best label to use for a fluorescence-activated cell sorting procedure? Assume that the total cell surface, target protein concentration is 10^{-14} mol/lit and the product of fraction labeled and fluorescent intensity must be greater than or equal to 1.5 in order to differentiate the desired cells from others.

(c) What would C have to cost for it to be as cost effective as B.

LABEL	DISSOCIATION CONSTANT (M)	FLUORESENCE INTENSITY (MAX = 10)	COST $US/MOL
A	10^{-6}	3	10^9
B	10^{-7}	9	10^{10}
C	10^{-8}	5	10^{11}

7.4 Your car has run out of gas. Therefore, its speed declines with distance x as you coast down the highway. Assume that the velocity v (miles/h) depends on distance past the point where you ran out of gas, according to

$$v(x) = 66 - 33\,x.$$

If the gas station is 1.8 miles away, how many minutes will it take to get there?

7.5 What does Equation 7.4b become if 99% conversion is sought?

7.6 If the same amount of enzyme is used, how much longer (e.g., twice as long) does it take for 99 percent of a substrate to be reacted to product as opposed to ninety five percent for the cases below?

(a) The dissociation constant and initial substrate concentration are 10^{-5} M and 10^{-3} M, respectively.

(b) The dissociation constant and initial substrate concentration are 10^{-5} M and 10^{-7} M, respectively.

(c) Why is there such a difference in the time factor when only S_o and not the percent conversion was changed?

7.7 An enzyme has the following properties: $V_M = 10^{-3}$ mol lit^{-1} s^{-1} and $K = 10^{-5}$ M.

(a) Is the K value an association or dissociation constant and why?

(b) How long will it take for the substrate concentration (S) to decrease from 10^{-5} to 10^{-6} M?

(c) If the initial product concentration is 0 M, what is the product concentration after the time found in part (b) has elapsed?

(d) By what factor (half, one hundred, etc.) does the time change when the enzyme concentration is doubled?

(e) By what factor (half, one hundred, etc.) does the turnover number have to change when the enzyme concentration is fixed, yet the reaction time is desired to be reduced by a factor of ten?

7.8 You are considering the purchase of two different automobiles with a zero interest loan. For the different scenarios below, pose an objective function and select which auto is better.

(a) Auto A costs $15,000 and the fuel costs are estimated to be $1,100 per year. Auto B costs $25,000 and the fuel costs are estimated to be $1,500 per year. Both autos and the loans will last ten years and have similar insurance and repair costs based on your excellent driving record and ability to fix things yourself.

(b) Auto A costs $15,000 and the fuel costs are estimated to be $1,100 per year. This auto (and simple loan) will last five years and Yugos tend to have zero resale value, but you will buy another Yugo because you think that they are cheap and all that you need. Auto B costs $25,000 and the fuel costs are estimated to be $1,500 per year over the ten-year lifetime of the car and simple loan. Again, there will be similar insurance and repair costs.

7.9 For an initial analyte concentration equal to 10^{-4} M, choose the best enzyme, based on the data below, to use in a sensor that converts the analyte to a detectable product, Assume that converting 95 percent of the analyte is acceptable for signal generation.

The shelf life of the enzyme-containing sensor component is one year.

Enzyme A: $K = 10^{-5}$ M, turnover number $= 10^4$ s^{-1}, and half-life $= 6$ months.

Enzyme B: $K = 10^{-6}$ M, turnover number $= 10^4$ s^{-1}, and half-life $= 3$ months.

7.10 Consider enzyme-based sensor design when S_o is much less than K.

(a) Show that ϕ is given approximately by the following simplified proportional relationship for a shelf life of one year:

$$\phi \sim (K/k_o)\ exp\ (1/\tau_H), \text{ where } \tau_H \text{ has units of years.}$$

(b) Using this relationship may simplify how to decide what enzyme alternative is best under some circumstances. For example, based on the properties of the two enzymes below, pick the one best for a sensor application when the shelf life is one year.

Enzyme A: $K = 10^{-5}$ M, turnover number $= 10^4$ s^{-1}, and half-life $= 6$ months.

Enzyme B: $K = 10^{-6}$ M, turnover number $= 10^4$ s^{-1}, and half-life $= 3$ months.

CHAPTER 8

Metabolic and Tissue Engineering and Bioinformatics

8.1 | Purpose of This Chapter

The prior chapter illustrated some of the technologies and bioengineering problems that are inspired by the attributes of biomolecules. This chapter will define and illustrate through examples the two prominent bioengineering domains wherein *intact* cells are manipulated for technological or medical purposes. The practitioners call these fields **metabolic engineering** and **tissue engineering**. Additionally, how information is gathered from cells and processed will be introduced. The sequencing of the genomes of numerous organisms has generated a need to understand how genetic assets and gene-based events collectively confer behavioral traits. From the technological standpoint, an improved understanding of how gene circuits function will improve outcomes when circuit manipulation is done deliberately to coax a microbe into producing a valuable molecule or when cells are being guided to grow into a functional tissue. **Bioinformatics** is the catchall term for generating and dealing with information from biological systems.

As noted in Chapter 0, engineering at the cellular level represents a blurring of the distinctions between some biochemical and biomedical engineers. Because product- or medically-inspired engineering at the cellular level both demand an understanding of cell function and behavior at the gene expression level, the blurring has increased. Therefore, it is useful to include both tissue and metabolic engineering in the same chapter. Including an overview of some aspects of bioinformatics in this chapter will shed light on developments aimed at better understanding and improving the outcomes of metabolic and tissue engineering.

After providing examples of recent metabolic and tissue engineering efforts, this chapter concludes with some suggestions on how to direct your future study in order to gain a deeper knowledge of the scientific and engineering fundamentals involved with cellular engineering.

8.2 | Microbial Metabolic Engineering

Basis and aims.   Microbes have been used for many years to produce wine, beer, cheese, antibiotics, and numerous other products that we now take for granted. Demonstrating in the 1970s that genes from animals or other sources can be "transplanted" into and expressed by bacteria opened many new doors for scientists and engineers.

First, the possibility emerged for using bacteria and yeast as "factories" for producing useful proteins and other molecules with therapeutic or commercial value. This possibility is intriguing because some bacteria can produce a vast array of biochemicals from inexpensive and simple materials such as glucose and salts. Moreover, the naturally occurring molecules that chemists find difficult to synthesize could instead be produced via microbes using simple raw materials.

The second and allied possibility that emerged was that it may now be possible to alter the regulation of metabolic pathways and gene expression such that the production of new products by cells could be made more efficient. That is, not only could new capabilities be imparted to some cellular "factories", but also the efficiency in how resources are directed to the synthesis of the new products could conceivably be enhanced. Alternately, many cells naturally produce many useful molecules such as amino acids. Therefore, it may be possible to increase the efficiency at which raw materials are directed to building the molecules many cells normally produce.

Metabolic engineering is the field where the genetic potential and regulation within cells are optimized such that the synthesis of a target product is maximized and the production of unwanted products is minimized. Regulation is managed through genetic manipulation, the selection of growth conditions, and the design of cell cultivation systems (**bioreactors**). The products cells normally produce and nonnative molecules encoded by foreign genes are both of interest.

A long history of metabolic engineering exists. In the past, mutated strains of bacteria, yeast, and fungi were generated and the few that exhibited a heightened ability to produce a useful molecule, such as the antibiotic penicillin, were isolated. This "hunting for a needle in a haystack process" is time- and labor-intensive. Using molecular biology tools now provides a more direct route to developing and isolating a "cellular factory" with high potential.

The enterprise of metabolic engineering involves both theoretical analysis and experimentation; hence, the field engages scientists and engineers with varied skills and interests. The use of theory (e.g., network traffic flow analysis) allows for the prediction of yield and pinpoints potential strategies for controlling the flow of resources within the cell. Experimentation involves genetic manipulation and the identification of how raw materials are actually routed through a cell's metabolic networks. Overall, the challenge a metabolic engineer faces is to identify an effective way to direct resources to the synthesis of the target product without causing the cellular factory to flounder excessively. Indeed, if a microbe or yeast is altered excessively, then the resulting inability to support basic energetic and replication functions may cause the whole process to fail or proceed at an unacceptably low rate.

Cell engineering example: Adding new genetic capability.   In Chapter 0, some enzyme-catalyzed reactions that can be performed *in vitro* with DNA were outlined. These reactions and other processes can be used to instill a new genetic-based capability in a cell. Metabolic engineers, and as will be seen at the end of the chapter, some tissue engineers, employ these methods as they attempt to improve existing technologies or explore new possibilities. When bacterial cells are coaxed to internalize and use the instructions provided by a piece of foreign DNA, the process is called **bacterial transformation**.

Using bacteria to produce a protein such as insulin based on human genetic instructions has many advantages. Prior to gaining the means to produce human proteins in microbes, insulin was obtained from animals that produce a similar protein. Insulin from pigs and cows differ from the human molecule by one and three amino acids, respectively. Although the animal-derived insulin substitute works, the differences between the human and animal insulin molecules can result in immune system activation. One adverse consequence is a higher dose of the animal-derived insulin is required to offset the effort the immune system exerts on removing the "foreign" molecule from the body. Bacteria such as *E. coli* also work pretty cheaply; only inexpensive and basic raw materials are required.

How bacteria can be transformed to produce human insulin and other nonnative proteins will be outlined. If the starting point is an isolated portion of human DNA that encodes for insulin, then the polymerase chain reaction method outlined in Chapter 0 can be used to increase the amount of DNA. With a lot of DNA now in hand, two routes are possible for introducing and expressing the insulin gene within bacteria.

One method is simple in procedure, but often many pitfalls must be solved before success results. At a particular stage in growth, bacteria such as *E. coli* have been found to be maximally receptive to importing DNA. These are called **competent cells**. The idea is that the DNA will enter the cell and a DNA recombination event will occur such that the human DNA is now integrated within the *E. coli* DNA.

While simple in concept, often a lot of work and trial and error is needed to insure that recombination occurs. The frequency at which the recombination event occurs can be low, so one has to look at many candidate cells to establish whether the desired events, recombination and insulin expression, have occurred. To improve the odds that cells can be found that indeed integrated new DNA, additional DNA that confers a trait such as antibiotic resistance may be linked to the insulin-encoding DNA. In this case, cells that grow in the presence of antibiotics may also have successfully incorporated the insulin-encoding DNA.

Another practical issue is that one has to make sure that the DNA fragment does not insert in the midst of a gene that is important for *E. coli*'s survival under the conditions that will be used to manufacture insulin. This problem can be solved by exploring the DNA sequence of *E. coli* and engineering the DNA fragment such that the odds are maximized that recombination will not occur in a vital section of the bacteria's chromosome. Finally, a regulatory sequence is often useful to incorporate, so that the newly added gene can be turned on and off.

Another strategy that requires more steps involves using **plasmids**. Plasmids naturally occur in bacterial cells. They are circular double-stranded DNA molecules that are much smaller than the chromosome. Depending on the plasmid, a few or hundreds of copies can exist in one bacterial cell. Plasmids confer specialized traits to bacteria, such as antibiotic resistance, and bacteria can exchange with each other these small repositories of genetic information. Because bacteria normally have plasmids and many are able to "absorb" or exchange plasmids, it is not surprising that genetic engineers have learned how to achieve their goals by piggybacking on the natural processes that involve plasmids.

When plasmids are used, the process starts as before. PCR is used to amplify the DNA that contains the information for producing insulin. The next steps, which are outlined in Figure 8.1, differ. A plasmid normally found in *E. coli* is cut with restriction enzymes (Chapter 0). Recall that such cutting results in the formation of "sticky ends." If the insulin DNA has been processed to have complementary sticky ends, then it will insert into the plasmid via base-pairing. Gaps in the single strands can be closed using DNA ligase.

Starting with many plasmid molecules and amplified DNA, many copies of a modified plasmid can be produced that also contain a **marker** for a trait that will be easy to identify (e.g., antibiotic resistance) and should be manifested by cells that have the plasmid within them. To introduce the plasmid inside the bacterial cells, **electroporation** can be used. Cells and plasmid DNA are inserted between two electrodes. A voltage is typically applied across a millimeter-scale gap, and the field gradient is significant enough to disrupt the cells' plasma membrane. Openings are produced in the membrane through which plasmid DNA can enter some of the cells. Some cells also do not survive or recover. Recalling Chapter 4, the "effect" of a field is related to the magnitude of the gradient; hence, the yield of transformed cells is linked to the voltage and gap size as well as the voltage pulse duration used. Too low of a gradient will result in a low transformation yield. When the gradient is too large, irreversible damage to the cells may occur, which will in turn lower the total number of transformed cells found.

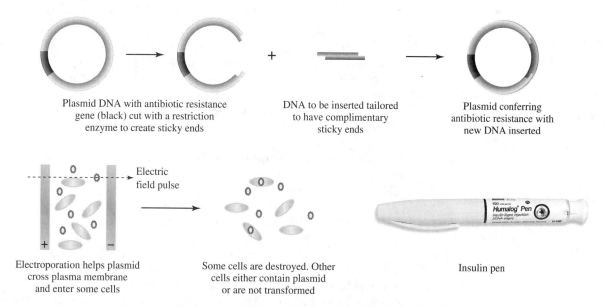

Plasmid DNA with antibiotic resistance gene (black) cut with a restriction enzyme to create sticky ends

DNA to be inserted tailored to have complimentary sticky ends

Plasmid conferring antibiotic resistance with new DNA inserted

Electric field pulse

Electroporation helps plasmid cross plasma membrane and enter some cells

Some cells are destroyed. Other cells either contain plasmid or are not transformed

Insulin pen

FIGURE 8.1 Schematic of how bacterial cells could be transformed to produce insulin and other products. After transformation, biochemical engineers grow the microbes and isolate the product. Ultimately, a product such as a packaged device for injecting insulin can result. (Device image courtesy of Eli Lilly and Company)

TABLE 8.1

Examples of Products Produced from Transformed Microbes	
Product	Medical Use
Insulin	Diabetes management
Factor VIII (see Chapter 11)	Treatment for hemophilia A
Factor IX	Treatment for hemophilia B
Human Growth Hormone	Treatment of dwarfism
Erythropoietin	Treatment of anemia
Tissue Plasminogen Activator	Blot clot dissolution (see Chapter 11)
Interferon	Augment immune system function

By using antibiotic selection, some of the successfully transformed cells can be isolated. In many cases, this is the end of the process; namely, *E. coli* has incorporated the instructions for synthesizing new protein. In other cases, *E. coli* is simply used to amplify the DNA further via growing a quantity of plasmid-containing cells. After processing the amplified DNA isolated from *E. coli*, the DNA may be used for transforming another type of cell. Chromosomal integration is a possible endpoint for plasmid-based strategies as well. Such integration, if it does not knock out an essential gene, is advantageous because the new DNA is permanently imbedded in the genome as opposed to being associated with a peripheral plasmid.

The transformed bacteria are grown in specialized vessels called **bioreactors** or **fermenters**. An example of a large bioreactor was shown in Chapter 0. The growth vessel and solution of nutrients (**growth medium**) are first sterilized. Thereafter, a starter culture of transformed cells is added. As the cells grow, they are supplied with oxygen and nutrients to foster their growth and to manage their metabolism such that the recombinant gene product is produced at a high level. The use of recombinant DNA technology has yielded a number of microbial-produced therapeutics.

The use of recombinant DNA technology has yielded a number of microbial-produced therapeutics. Partial listing of products is provided in Table 8.1, and Figure 8.1 shows an example of where imparting new capabilities to microbes can lead: the development of pre-filled devices for dispensing human insulin.

Metabolic engineering example: Lowering cellular energy use.   An interesting and early metabolic engineering experiment that used modern molecular biology tools involved lowering the energetic (ATP) cost of cell production. Because energy is derived from the oxidation of carbon compounds and these energy sources cost real money, lowering the energetic cost of producing cells means that less raw materials and money are needed to produce a certain quantity of cells. Raising the yield Y (see Chapter 1) is technologically and economically akin to improving airplane fuel efficiency and thus the profitability of an airline.

The aim was to lower the ATP costs associated with converting the abundant nitrogen source, ammonium ion (NH_4^+), to a key amino acid, glutamate (see Figure 1.3), in the bacterium *Methylophilus methylotrophus*. Glutamate is a prevalent constituent of cellular proteins; hence, the net ATP required to build cell mass could be reduced if less ATP was needed to produce glutamate. *Methylophilus methylotrophus*, in turn, grows on methanol. It is grown in large-scale processes and the cellular proteins have been harvested and used as a source of animal feed. Thus, it would be advantageous to produce more protein (i.e., cell mass) per amount of methanol used.

In *Methylophilus methylotrophus*, the reactions used to convert ammonium to the amino acids glutamine and glutamate are*

$$(1) \text{ glutamate} + NH_4^+ + ATP \rightarrow \text{glutamine} + ADP + Pi$$
$$\underline{(2) \text{ glutamine} + 2\,\alpha\text{-ketoglutarate} + NAD(P)H \rightarrow 2 \text{ glutamate} + NAD(P)^+}$$
$$\text{NET: } NH_4^+ + \alpha\text{-ketoglutarate} + ATP + NAD(P)H \rightarrow$$
$$\text{glutamate} + ADP + Pi + NAD(P)^+$$

The net reaction that forms glutamate "costs" one ATP per mole of glutamate. In *E. coli*, a different enzyme catalyzes a reaction that yields glutamate without using ATP; hence, the *E. coli* enzyme potentially provides an energetically "cheaper" way to produce glutamate. The reaction that occurs in *E. coli* is

$$NH_4^+ + \alpha\text{-ketoglutarate} + NAD(P)H \rightarrow \text{glutamate} + NAD(P)^+.$$

To evaluate the potentially "cheaper" route, recombinant DNA techniques were used to essentially replace reaction (2) with the reaction catalyzed in *E. coli*. In *Methylophilus methylotrophus*, the gene that encodes for the enzyme that catalyses reaction (2) was "silenced", and the gene that encodes for the *E. coli* enzyme was inserted (Windass, J.D., Worsey, M.J., Pioli, E.M., Pioli, D., Barth, P.T., Atherton, K.T., Dart, E.C., Byrom, D., Powell, K., and Senior, P.J., 1980). Consequently, all the glutamate that *Methylophilus methylotrophus* requires could be synthesized without paying an ATP tax. Some ATP would still be required, however, to synthesize the glutamine needed via reaction (1).

The modified organism converted 4 to 7 percent more of the carbon in methanol to cell mass (i.e., protein) than the original organism. While the enhancement in yield may not sound dramatic, even small gains can have a positive economic impact in a large-scale process or system. Indeed, in the large United States economy, a less than one percent change in interest rates can have a significant overall impact on lending and spending.

Metabolic engineering example: Lowering waste. One important problem in using microorganisms as "factories" for producing products such as recombinant proteins is that inexpensive carbon and energy sources (e.g., glucose) are not 100 percent used to form CO_2, cells, and valuable biological molecules. Rather, up to 40 percent of the glucose carbon can be partially broken down to waste products such as acetic acid. From the engineering standpoint, the production of acetic acid versus the desired product represents a waste of raw materials and raises the production cost. Apart from economics, the accumulation of acid waste products can result in growth rate reduction and other toxic effects.

There are a number of ideas for why glucose is not completely used. One interesting idea is that microbes partially break down carbohydrates and then

---

*In Chapter 4 (Fig.4.2) NAD^+ and NADH were introduced. $NAD(P)^+$ denotes that NAD^+ and/or the phosphate-bearing form ($NADP^+$; used in biosynthetic metabolism) can participate in the reaction.

"store" some fraction in the growth medium for later use. That is, glucose is partially degraded to acetic acid. Thereafter, if the acid concentration is not at a toxic level, once the glucose is gone, the acetic acid can then be metabolized to sustain the growth of the microbial population, albeit at a lower rate. Such storage in the medium may represent an efficient strategy to the cells, but to the engineer, the carbon from the glucose is wasted and the kinetics is complicated and extended.

Many metabolic engineers have examined the carbon processing pathways in bacteria, along with their regulation, for the purpose of identifying how carbon wastage can be reduced or eliminated. The idea is that the carbon processing reactions used and the feedback regulation mechanisms can be altered by manipulating the genes that encode for the different enzymes involved.

One example is shown in Figure 8.2, which illustrates the metabolic subsystems involved with processing glucose and a hypothesis for how acetic acid formation occurs. Glucose is imported into a cell through the action of the **phosphoenol transferase system (PTS)**. What happens is that glucose reacts with a phosphorous-containing metabolite, called phosphoenol pyruvate (PEP), as follows:

$$Glucose + PEP \rightarrow Glucose\text{-}P + Pyruvate.$$

The important outcome is that glucose acquires a phosphate group (PO_4^{3-}) and thus a negative charge. Acquiring this charge helps to trap glucose in the cell because the egress of a water soluble, charged chemical species through the hydrophobic, lipid (i.e., fatty) cell membrane is not favorable.

Once glucose has been phosphorylated upon its entry into the cell, it is broken down by a metabolic subsystem called **glycolysis**, which was also introduced in Chapter 3 (Figure 3.4). Here, glucose, which is composed of six carbons $(C_6(H_2O)_6)$, is split into two, three-carbon molecules called pyruvate. Note that pyruvate is also formed by the PTS. As glucose is broken down to pyruvate, some oxidations occur and the energy is stored in the ATP molecule. Some carbon is also drained along the way to feed biosynthetic processes. One example is the **Hexose Monophosphate (HMP) Pathway**. Early in the processing of glucose, some carbon enters the HMP pathway, where reactions convert the glucose-carbon into the monomers that will later be assembled into RNA molecules and DNA.

The end-product of glycolysis, pyruvate, then enters the **tricarboxylic acid (TCA) cycle**. The cycle gets its name from the chemical nature of the metabolites found in the TCA cycle. (Three carboxyl moieties (Chapter 2) can be found on one of the early molecules.) From the functional standpoint, additional oxidations occur in the TCA cycle that lead to ATP formation by oxidative phosphorylation. (Recall NADH oxidation in Chapter 4.) Some TCA metabolites are also bled out for biosynthesis. Indeed, the carbon skeleton for glutamate (see Figure 1.3), α-ketoglutarate, is produced in the TCA cycle.

Acid formation may occur because glycolysis proceeds faster than the TCA cycle can process pyruvate. That is, cells are engineered well by nature to remove glucose from the environment, but they are regulated or limited in the ability to fully use glucose. Thus, the reaction

$$Pyruvate \rightarrow Acetic\ Acid + CO_2$$

has an opportunity to occur, which leads to the accumulation of acetic acid (and other acids) in the growth environment.

Central carbon metabolism

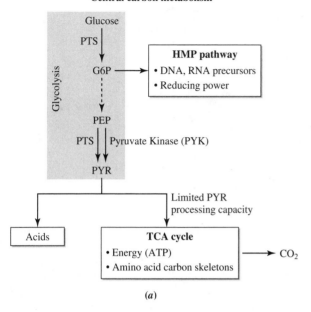

(a)

Carbon overflow eliminated by reducing PYK & PFK activities

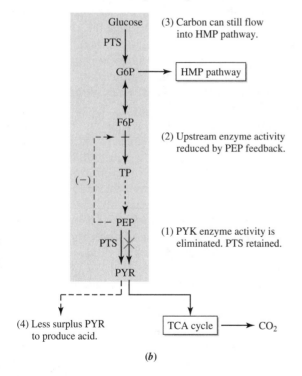

(b)

FIGURE 8.2 The relevant metabolic pathways and regulation associated with acetic acid by-product formation and by-product reduction. (a) How glucose is degraded for energy capture and precursor synthesis. (b) Eliminating the enzyme, pyruvate kinase (PYK), in tandem with existing feedback regulation may prevent by-product formation. (c) The elimination of PYK significantly reduces acid formation compared to the wild-type cell. Abbreviations: PTS, phosphoenol transferase system; G6P, glucose-6-phosphate; F6P, fructose-6-phosphate; TP, triose phosphate; PEP, phosphoenol pyruvate; PYR, pyruvate; TCA, tricarboxylic acid; and HMP, hexose monophosphate.

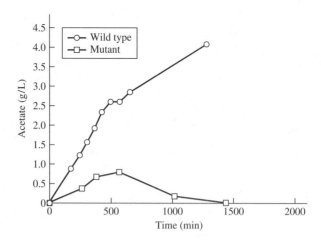

FIGURE 8.2   (*Continued*)

What are some ways to prevent acetic acid formation? One possibility that may occur to you is to mutate the PTS system so that the cell's ability to import and pass glucose through glycolysis is reduced. Reducing the importation rate may result in the TCA cycle not being swamped, which, in turn, would leave less pyruvate remaining to react to acetate. This strategy has been shown to work, but there is also a potential trade-off. Restricting glucose uptake would also potentially starve the side reactions that siphon off metabolites for other purposes (e.g., the HMP pathway). Thus, acetic acid by-products might be reduced, but at the cost of disrupting other processes, with the result that the cellular growth rate is diminished. This illustrates that the metabolic engineer, like other engineers, approaches problems at the systems level and has to weigh the trade-offs associated with a design strategy.

To provide some closure on this problem, Figure 8.2(b) shows one strategy that worked in the bacterium *E. coli* in that acetic acid formation was significantly reduced while the growth rate of the engineered cells was not substantially affected. A two-part strategy was used. The first part involved reducing the formation of the acetic acid precursor, pyruvate. There are two enzyme-catalyzed reactions that lead to the formation of pyruvate. One reaction has already been introduced: the PTS system. The second reaction is the step in glycolysis where **pyruvate kinase** (**PYK**) also catalyzes the conversion of PEP to pyruvate. Therefore, to lower the amount of pyruvate produced and to lessen the overwhelming of the TCA cycle, the PYK enzyme was "removed" from *E. coli* (Zhu, T., Phalakornkule, C., Koepsel, R.R., Domach, M.M., and Ataai, M.M., 2001) and another organism used in biotechnology, *Bacillus subtilis* (Fry, B., Zhu, T., Koepsel, R.R., Phalakornkule, C., Domach, M.M., and Ataai, M.M., 2001). Removal was accomplished by using recombinant DNA techniques akin to those presented earlier. The method entailed inserting a "junk" piece of DNA into the gene that encodes for PYK; hence, when the mRNA from the disrupted gene was translated, a nonfunctional PYK resulted.

At this juncture, glucose is free to enter the cell because the PTS system was not altered. Therefore, the HMP pathway presumably will not be starved. However, because PEP may accumulate due to the loss of the enzyme that removes it (PYK),

PEP accumulation has to be checked. The second part of the strategy entailed tapping into a naturally present feedback loop. PEP can bind to an upstream enzyme and slow its formation as shown in Figure 8.2(b). Thus, excessive accumulation of PEP is checked, but phosphorylated glucose is still free to enter the HMP pathway.

The effect of eliminating of the PYK enzyme on acid formation is shown in Figure 8.2(c). Compared to **wild-type** (i.e., before metabolic engineering) *E. coli*, the metabolically engineered *E. coli* produced considerably less acetic acid. The growth rate (not shown) was not significantly altered.

8.3 | Tissue Engineering

Motivation for tissue engineering.   Traditionally, four approaches have been used to treat the problems created by damaged or failing organs and tissues. First, drugs can be used to lessen symptoms or to compensate for organ failure. Insulin injections, for example, can be used to compensate for diminished pancreas function. Secondly, sometimes tissue can be borrowed from one part of the body to treat a problem elsewhere. Harvesting skin from an unburned region to rebuild a burned area is one example of what is called **autografting**. However, finding a source of the tissue on the body for use elsewhere can often be problematic. For example, we have only one heart, so it is impossible to procure a heart or even a valve to make a repair. When vessels from the legs or elsewhere are harvested and used to treat coronary artery disease, it may not be feasible to go back to the source again if the transplanted vessels later fail, because little material remains to harvest, or leg function will be unacceptably compromised if more vessels are removed for transplantation elsewhere. Thus, the third approach is to use tissues and organs that other people donate. When the tissue is derived from a source other than the recipient, it is called an **allograft**. To keep the tissue grafting terminology straight, recall an allosteric enzyme possesses an "other" site; hence, an allograft comes from a source "other" than the patient.

There are numerous challenges with using allografts. Availability is chronically less than demand. Moreover, foreign tissues and organs are prone to rejection by the recipient's immune system. Consequently, many transplant candidates succumb while waiting for a suitable donor. Those that are the lucky recipients of successful allografts often must take immune system suppression drugs for the rest of their lives, which can have side-effects.

The fourth approach is to construct and implant an artificial version of the damaged organ. The total replacement, mechanical heart is one example that will be discussed in Chapter 12. Many strides have been made, but totally artificial replacements still fall short of the performance of natural organs.

Instead of the four alternatives above, tissue engineers envision that diseased tissues and organs can be replaced by coaxing primitive cells or other cells into growing into a functional replacement. The definition and organization of tissue engineering research has benefited from specialists sharing views and needs at workshops sponsored by the National Science Foundation. (An updated report can be found at http://www.wtec.org/loyola/te/.)

Tissue engineering strategies.   From the biological standpoint, there are three basic approaches under investigation in tissue engineering research and development. One entails starting with primitive human cells that contain a full set of instructions, but yet have not executed many of the instructions by growing and forming a functional assembly of cells. By applying knowledge of how signal molecules allow for the instructions to be used along with physical support structures called **scaffolds** or **matrices** to guide the construction of a three-dimensional tissue, researchers hope to use undifferentiated cells as a starting point for regenerating functional tissue replacements.

The second approach involves expanding more mature cells to produce a reasonable facsimile of a functional organ or tissue. To achieve this goal, scaffolds are again designed and used that steer the cells toward attaining a mass with the correct geometry and size, while still presenting the metabolic and other capabilities required for the assembly to assume the needed function within the body after implantation. Typically, the scaffold is designed to disintegrate during growth so that only an organized collection of cells remains.

The third strategy consists of using already functional cells, tissues, and organs from animal sources. To prevent rejection by the human recipient, it is envisioned that animals can be genetically engineered to not produce molecular signatures that engage the human immune system. Alternately, methods can be developed to protect the foreign cells from the human immune system. An example is the use of implanted porcine (pig) islet cells. The insulin produced by porcine islet cells is active in humans. Thus, some researchers envision that implanted porcine islet cells can be used to rectify diabetes in humans. One major challenge, however, is to keep the foreign islet cells alive and protected from the human immune system while still allowing for the correct amount of insulin to be produced and released to the body. Membrane encapsulation is among the approaches that have been investigated. The concept entails putting the porcine islet cells in small porous "bags." If the pore size is chosen correctly, then the large cells of the immune system cannot encounter and attack the islet cells, but the smaller insulin molecule can still escape.

Another way to organize tissue engineering is by how completely the construct biologically replaces the native tissue upon implantation. Skin replacement is an example of a nearly full replacement of the native tissue. A sheet of skin is first grown apart from the patient using donated infant foreskins as the starting point. The sheets of skin are then used as a graft. An alternate tissue engineering technology entails designing a physical scaffold that may also contain growth factors, signaling factors, and/or seed cells, and then implanting the construct into the patient to prompt cell growth and organization. Here, the technology is more "enabling" in intent as opposed to generating a highly functional replacement at the outset. The last variation is represented by hybrid constructs. Blood vessel replacements, for example, can be fabricated from synthetic and cellular materials prior to implantation. The synthetic materials endow the replacement with strength and other desirable physical properties. However, growing blood vessel cells to line the inner, blood-contacting surface of the replacement vessel potentially provides a surface that is less reactive with blood and the body's immune system. Therefore, the construct attempts to exploit the desirable features of biological and synthetic materials.

From these attempts at classification of technologies and approaches, it should be apparent that tissue engineering combines a fundamental understanding of cell growth kinetics, regulation/differentiation, and materials engineering. How cells adhere to scaffold materials, migrate, and respond to scaffold-associated and medium factors must be understood. Finally, cell to cell interactions must be managed to create functional tissues. The scaffold-cell paradigm is an omnipresent feature of tissue engineering, and it should be apparent that the mechanisms of ligand-protein binding, which were discussed in Chapters 5 and 6, underlie the phenomena of cell migration, signaling, gene induction, etc. Because of the numerous complexities and number of cellular actors, the scientific and engineering base of tissue engineering fundamentals is not as developed as is the case for metabolic engineering; hence, many problems await solution by established and new investigators with fresh ideas.

Because tissues contain a variety of cell types and blood vessels, it can be challenging to achieve a fully functional structure that *also* readily integrates with the blood and other circulatory systems in the body. Thus, it is not surprising that most of the initial successes and current biomedical products have involved constructing thin tissues such as skin and cornea. Some of these successful examples will be described next to illustrate further the current elements of tissue engineering.

Tissue engineering example: Cornea replacement. In July 2000, excitement was generated by two different reports on the successful transplantation of engineered corneal tissue. Dr. Tsai and associates at Chang Gung Memorial Hospital and the University of Taoyuan, Taiwan, reported the successful improvement of the vision of six transplant recipients (Tsai, Ray Jui-Fang, Li, Lien-Min, and Chen, Jan-Kan, 2000). The second concurrent report was made by Drs. Ivan Schwab and R. Rivkah Isseroff (see http://news.ucdmc.ucdavis.edu/Isseroff_Schwab.html) of the University of California at Davis, School of Medicine and Medical Center (Schwab, I.R., Reyes, M., and Isseroff, R.R., 2000).

Corneal transplantation was (and remains) an important tissue engineering target for several reasons. There are over 45,000 transplants performed per year in the US to remedy age-, disease-, and injury-induced vision loss. Allografts work well because the cornea, which is the outer "skin" and refractive element of the eye's light-gathering section, is not very accessible to the body's immune system. Thus, modest and local versus radical, systemic immune suppression via drugs is required for successful transplantation of a corneal allograft. However, infectious diseases of the eye and injuries are extensive worldwide. Indeed, infections that lead to blindness are estimated to total six million per year (http://grants.nih.gov/grants/guide/pa-files/PA-02-053.html). Another factor to consider, according to the National Institutes of Health (http://grants.nih.gov/grants/guide/pa-files/PA-02-053.html), is that future donor sources may actually decline to the growth and successful implementation of another medical technology. Patients who have had their vision corrected by laser *in situ* keratomileusis (LASIK) cannot donate usable corneal tissue despite their own vision being improved. Thus, many people could benefit from an easier and low risk means of vision restoration that does not depend on donors.

The University of California at Davis team used a four-step process, which is summarized in Figure 8.3. First, presumed corneal stem cells were isolated from the patient's healthy cornea. If the corneas of both eyes were damaged or diseased, then corneal stem cells were isolated from a relative of the patient. Both sources of stem cells were used in the trial conducted by the team at the University of California at Davis. Secondly, the stem cells were divided and proliferated in different culture

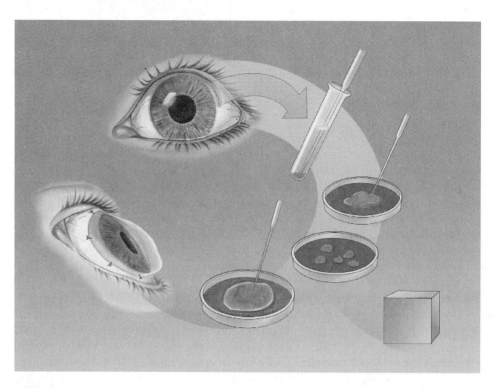

FIGURE 8.3 The steps in the procedure the University of California at Davis team used to generate replacement corneal tissue. Presumed corneal stem cells are harvested from the patient's healthy eye or a related donor. The tissue is separated for growth in laboratory dishes, where the stem cells produce a layer of cells one cell thick. The expanded cells including the surviving stem cells, are transferred to the surface of an amniotic membrane. Some cells also can be frozen and banked for later use. On the membrane, the corneal cells form a layer 5 to 10 cells thick. A tissue that combines the elasticity and resilience of amniotic membrane with the biological properties of corneal tissue is yielded. This bioengineered composite tissue is then stitched onto the patient's eye, after removing the abnormal corneal tissue. (Figure and process summary are courtesy of the University of California at Davis Medical Center.)

dishes to generate a monolayer of cells, some of which were the original stem cells. Some of the cells from the multiple dishes were stored for later use, in case the transplantation did not work and the procedure had to be repeated. Thirdly, the expanded cells were transferred to an amniotic membrane. This membrane material, which functions as a scaffold in this case, is obtained from mothers' afterbirth. On the membrane, the expanded cells grew further to form a layer that was five to ten cells thick. This thicker layer of cells apparently acquires some of the properties of a healthy cornea, which is characterized by a particular cellular organization and refractive (light-bending) properties. Lastly, the thick layer of cells was transplanted to eyes that had abnormal tissue removed beforehand.

    This success has generated much follow-up research to better understand the process and to improve the outcome. For example, a normal cornea is composed of three cellular layers. The outer thin layer is comprised of epithelial cells. Beneath that layer resides a thicker layer of connective tissue that contains, in part, dormant fibroblasts. Finally, another thin layer of epithelial cells exists that is in contact with the eye's interior fluids. Researchers are now interested in better replicating the middle layer to acquire the correct combination of mechanical strength and optical properties. An example of ongoing work entails developing matrices/scaffolds that are an alternative to amniotic membranes. Protein-based gels are among the alternatives under investigation. Isseroff and colleagues have recently reported on using cross-linked **fibrin gels** as a template for corneal stem cell proliferation (Han, B., Schwab, I.R., Madsen, T.K., and Isseroff, R.R., 2002). As will be discussed in Chapter 12, fibrin is the major structural component of blood clots. When cross-linked, effectively a porous array of interconnected fibers is created.

Tissue engineering example: In situ disease rectification.   Earlier it was noted that the efforts of metabolic and tissue engineers can overlap in that they both work with intact cells and genetics, and regulatory mechanisms are managed to achieve a desired goal. The overlap has been extended by recent tissue engineering research that entails implanting scaffolds and cells not solely for the purpose of re-placing diseased or injured tissue. Rather, the implanted cells are intended to inte-grate with host tissue *and*, because of their modified genetic "programming," substances will be produced that may alter the course of a disease. For example, the integrated and engineered cells could produce a substance that was missing or inac-tive beforehand due to a genetic defect. Alternately, the integrated and engineered cells could be envisioned to produce a substance that slows or stops the proliferation of cancer.

One recent example that has been investigated in animal systems illustrates the approach and aims. However, some additional biology needs to be first introduced. There are two ducts in embryos that develop into female reproductive organs, or disappear in males while leaving only a **vestigial** (i.e., remnant) structure. These ducts bear the name of an early German investigator of embryo development, **Johannes Muller** (1801–1858), and are called **Mullerian ducts**. As you may expect by now, gene expression and chemical secretion events occur that direct the devel-opment process. **Mullerian Inhibiting Substance (MIS)** fosters regression of the ducts in the human embryo. MIS is actually a **glycoprotein** (a protein with carbohy-drates attached).

Interestingly, MIS promotes the regression of ovarian tumor cells; hence, some researchers view MIS as a potentially useful chemotherapy agent. However, purifi-cation of MIS and administration to a particular location within the body is not straightforward. Indeed, in Chapter 13, you will be introduced to pharamaco-kinetics, which is the analysis of how drugs and substances disperse in the body in a time-dependent fashion. A substance injected into the body experiences numerous fates, such as elimination, degradation, and binding to blood proteins. Therefore, attaining a therapeutic dose at a particular location while minimizing side-effects can be quite challenging.

A potential alternative is to genetically modify ovarian cells prior to implanta-tion into the body, such that MIS is produced. Such genetic modification is termed **transfection**. If provided a scaffold in the ovary, the transfected cells can expand and thus form a tissue that produces MIS, which in turn, might promote the regression of a tumor present in the reproductive organs or nearby.

A team of researchers from Harvard Medical School and Massachusetts General Hospital explored this strategy in mice with compromised immune systems (Stephen, A.E., Masaikos, P.T., Segev, D.L., Vacanti, J.P., Donahoe, P.K., and MacLaughlin, D.T., 2001). The scaffold they used was composed of **polyglycolic acid (PGA)**. Figure 8.4 shows an abbreviated pathway for how PGA is produced, and its repeating unit. PGA is a member of the **polyester** class of polymers. The "ester" portion of the name is derived, in part, from the bond pattern present. The **ester linkage** $[-(C=O)-O-C-]$ joins sections of the molecule together. One way to synthesize an ester is to conduct a **condensation reaction** between an **alcohol** $[R-CH_2-OH]$ and a **carboxylic acid** $[R'-(C=O)-OH]$. The reaction is called a condensation reaction because one molecule of water (H_2O) is eliminated

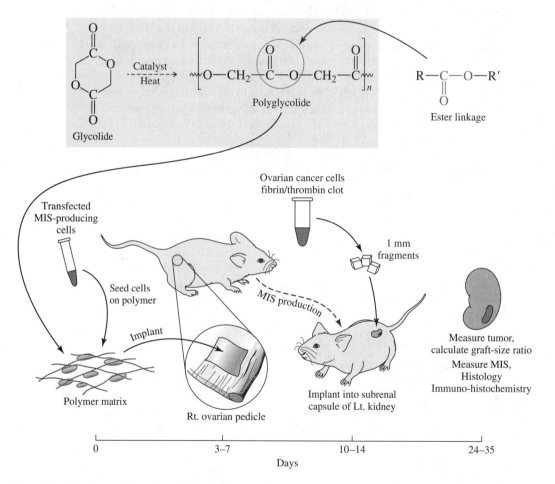

FIGURE 8.4 (Top) The chemistry of producing the polymer polyglycolic acid (PGA), which is a member of the polyester class of polymers. (Bottom) Method for implanting MIS-producing cells in mice and testing for antitumor efficacy. Initially, cells are seeded onto the polymer. Days afterward, the cell-impregnated polymer graft is implanted on the right ovarian pedicle. Thereafter, a human tumor fragment is implanted under the left renal capsule. Blood is collected throughout the protocol to measure MIS. MIS levels increased with the size of the size of implanted transfected cells. Evidence of tumor regression was observed in the mice implanted with transfected cells. (Figure adapted from Stephen, A.E., et al. *PNAS* 98 (2001): 3214–3219).

each time one alcohol and one carboxylic acid molecule combine:

$$R'—C—OH + R—OH \rightarrow R'—C—O—R + H_2O$$
$$\quad\, \| \qquad\qquad\qquad\qquad \|$$
$$\quad\, O \qquad\qquad\qquad\qquad O$$

A *poly*ester thus contains a repeating unit in which an ester linkage joins constituent R groups (i.e., $—[O—R'—(C{=}O)—O—R—(C{=}O)]_n—$). What can be synthesized by condensation can also eventually be reversed. **Hydrolysis** involves the breakage of an ester linkage through the addition of water to reverse the reaction written above, resulting in the production of alcohol and carboxylic acid fragments. The hydrolysis reaction is what provides for dissolution or reabsorption of polyester-based scaffolds after implantation.

PGA has been traditionally used for sutures, and it "dissolves" over weeks or months via hydrolysis when implanted in the body. (See http://www.ifsa.com.mx/apg.htm, http://www.devicelink.com/mpb/archive/98/03/002.html.) The long history of using PGA in medical applications is why some tissue engineers use PGA as a starting point when an absorbable, nontoxic scaffold material is desired. Other polymers are also used or in development for tissue engineering.

A schematic of the overall experimental procedure is provided in Figure 8.4. Initially, transfected cells were seeded and grown on a PGA scaffold. Different size implants were used on order to establish how the established MIS level depended on implant size. A correspondence was found between implant size and MIS level measured in the blood. Accordingly, a basis for choosing an implant size was established for the testing of antitumor efficacy. The efficacy was determined by implanting human tumor tissue into the mice and measuring subsequent tumor mass and tissue characteristics in mice that produced MIS and those that did not. The results were promising, in that tumor proliferation was statistically less in MIS-producing mice.

The MIS-producing cells that were implanted into the mice were derived from another animal source. The transfected and implanted cells were **Chinese Hamster Ovary (CHO) cells**. CHO cells are commonly used because much is known about how to transfect them. Because the immune systems of mice would normally attack CHO cells, mice with suppressed immune systems were used. Thus, many more steps must be taken before transfected cells can be routinely used for combating diseases in humans. However, the CHO-mouse-human tumor system does illustrate how our accumulating knowledge of cell biology can be tested and potentially harnessed.

8.4 | An Experimental Facet of Bioinformatics

Motivation.   As we have seen, metabolic and tissue engineers both harness the regulatory systems and genetic information within cells. In some applications, they also both try to "fool" or alter regulatory systems in order to get the result they desire. How do they know what genes are expressed in the first place? How does one know that a particular manipulation actually results in gene expression? How can the effect of expressing a gene that is naturally present (or introduced via transformation) on the other gene expression events be understood? That is, if one part of a complex circuit is "tweaked" or a new element is added, what is the effect on how the other parts of the system operate? Finally, how does the growth environment alter what genes are expressed and their expression levels? The answers to these questions are vital for making the technologies of metabolic and tissue engineering yield more predictable outcomes. The answers also increase our general understanding of cell function.

One obvious way to answer the above questions is to analyze the product(s) of gene expression. If, for example, a given gene encodes for an enzyme, then the total protein isolated from a cell can be analyzed to determine what enzyme activities are present. For example, if alcohol dehydrogenase activity is present (recall Figure 4.3), adding ethanol and NAD^+ to a cell protein sample should result in observable NADH formation. The rate at which NADH is formed, in turn, indicates the expressed level of alcohol dehydrogenase. This traditional procedure has contributed to our current understanding. However, the approach is labor intensive. A global view of how gene expression networks operate is also hard to construct with a single measurement approach.

DNA microarrays report gene expression.   Recall from Chapter 1 that when a gene is expressed, a working copy called messenger RNA (mRNA) is first produced. This working copy is then translated to produce a protein. Therefore, if a way could be devised to inventory the types and abundances of all the mRNAs present in a cell, then one could have a "snapshot" of what genes are currently expressed and roughly at what level. The latter assumes that the abundance of a particular mRNA corresponds to the level at which a particular gene is being expressed. Moreover, if a baseline condition is established for a particular environmental situation, then one can assess which genes are "up regulated" or "down regulated" when environmental conditions are altered.

The technology for obtaining this mRNA "snapshot" now exists. The technology has been given different names in scientific journals and the popular press. Common names are **DNA microarrays, biochips,** and **gene arrays.** Despite an abundance of names, similar operating principles apply. Different segments of sequence information from a cell's DNA are first attached to a surface such as a glass slide. In practice, a slide may contain on the order of one thousand different "spots" where a different DNA sequence is present at each. Because the DNA from many organisms has been (or soon shall be) sequenced, each segment can, for example, relate to a specific gene in a cell's genome.

The cells being subjected to analysis under a particular biological or environmental state contain many mRNAs of differing abundances. DNA copies of the cellular mRNAs can be made via the activity of the enzyme **reverse transcriptase.** A given DNA copy can bind to a surface-bound segment via base-pairing (recall Chapter 1) if the complementary surface-bound segment is present. The DNA copies are also labeled with a fluorescent dye. Thus, wherever a binding event occurs on the glass slide, a bright fluorescent spot will appear. Nonfluorescent spots indicate that there was no match between the surface-bound DNA and mRNA-derived, copy DNA.

In practice, two treatments are often applied to one DNA array. This means, for example, that DNA copies are made from mRNA present in cells grown under two different conditions. The DNA copies from the two different conditions are also labeled with different fluorescent dyes. For example, a bacterium such as *E. coli* is grown on two different carbon sources. When grown on one carbon source, the DNA copies of the mRNAs are labeled red. When grown on a different carbon source, the DNA copies of the mRNAs are labeled green. When the samples of red- and green-labeled DNA copies are mixed and applied to DNA arrays, there are four outcomes in terms of "spot" coloration: (1) no bright spots, (2) red spots, (3) green spots, and (4) spots that vary in yellow coloration. The first case indicates, for example, that a subset of genes is not expressed when either carbon source is metabolized. The second and third cases indicate that different subsets of genes are expressed when growing on one carbon source or the other. The last case indicates that some genes are coexpressed, but the ratio of expression may or may not vary. The method just outlined is used for **expression analysis** because it is the level of mRNA that is analyzed, albeit somewhat indirectly due to the intermediary mRNA $\rightarrow$ fluorescent DNA copy step.

Another important application is **genome typing**. DNA fragments are again first spotted on the glass slide. However, as opposed to investigating mRNA levels, genomic DNA fragments from a cell are directly used after they have been tagged with fluorescent molecules. Genome typing can be used, for example, to determine if an organism contains a gene similar to different, yet more fully characterized organism. A gene inventory can be developed in the less characterized organism because the DNA that codes for the same function (e.g., a structural gene for alcohol dehydrogenase, which yeast, humans, etc., possess) in different organisms can be highly homologous and thus some base-pairing can occur.

An example of the output from a DNA array experiment is shown in Figure 8.5. The spot intensities and ratios are measured with a scanner. Lasers are used to excite the fluorescence (recall laser excitation in fluorescence-activated cell sorting in Chapter 7) and the image is digitized. Digitization enables the calculation of intensity ratios. Numerous databases have been constructed where researchers archive their array results. The **National Center for Biotechnology Information** (NCBI) is one repository and distributor of genomic and expression data (http://www.ncbi.nlm.nih.gov/geo/info/print_stats.cgi) that is open to all to learn more, and the current data formats can be perused.

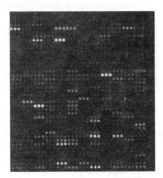

FIGURE 8.5   An example of the raw signals from a *section* of a large "gene chip" and how the data is processed. Different DNA sequences were first attached to the surface. Moreover, each DNA sequence was attached in triplicate to enable better averaging of signal intensity. Then, the mRNAs from two different lines of pancreatic cells were converted to complementary DNAs (cDNAs). The cDNAs derived from cell line 1 and 2 were also labeled with either a red or a green fluorescent dye, respectively, to distinguish the cDNAs according to their source. The intensity of the total fluorescence emitted from each location can be measured with a scanning device. The fluorescence signal has two components: red and green. The image shows the magnitude of the green fluorescence at each location in a section of the chip. Image courtesy of D.J. Hollingshead (Genomics and Proteomics Core Laboratories, University of Pittsburgh) and Dr. K. Poque-Geile (Department of Medicine and Center for Genomic Sciences, University of Pittsburgh, Pittsburgh, Pennsylvania).

8.5   Computational Component to Bioinformatics: Eigen Value-Based Methods

The problem of extracting information from DNA arrays.   There are many genes in a cell, and the collective state of expression of as many genes as possible is needed to acquire a complete picture of regulation. Consequently, gene array fabricators have responded by building chips with hundreds or thousands of spots. However, being presented with thousands of pieces of information can easily overwhelm the unaided brain. The result is that overall patterns and relationships cannot be readily discerned. Here, patterns mean, for example, that a subset of genes is always "ON" under the two conditions, which suggests that the same control system may link the expression of the genes in the subset. Alternately, under one set of conditions, some genes are expressed while others are not. When multiple time points are surveyed with DNA chips, how fast different genes turn on and off is gleaned as well as which subsets are related or unrelated to each other.

The rewards for having an effective means for processing the large amount of data from DNA arrays are manifold. The expression pattern can be related back to the environment, thereby providing a link between genotype and phenotype. Additionally, negative and positive control signals and circuits can be hypothesized based on the data, and in the process, a basis for subsequent experiments is defined.

Engineers, applied mathematicians, and others are striving to surmount the problem of being overwhelmed by a large amount of data. A number of methods now in use are based on **eigen value analysis**. Two popular methods are **singular value decomposition** and **principal component analysis**. The essence of these methods, eigen value analysis, is described next after presenting some background on matrix manipulation and the information provided in some matrices.

Information and correlation matrices.   Recall that the **multiplication of two matrices** occurs by element-wise, column-row multiplication and addition as follows:

$$\begin{pmatrix} A & B \\ C & D \end{pmatrix} \begin{pmatrix} E & F \\ G & H \end{pmatrix} = \begin{pmatrix} (AE + BG) & (AF + BH) \\ (CE + DG) & (CF + DH) \end{pmatrix}.$$

The **transpose** of matrix $\mathbf{A}$, which is denoted by $\mathbf{A}^T$, is formed by interchanging rows and columns. For example,

$$\mathbf{A} = \begin{pmatrix} 0 & 3 \\ 1 & 4 \\ 2 & 5 \end{pmatrix} \quad \text{and} \quad \mathbf{A}^T = \begin{pmatrix} 0 & 1 & 2 \\ 3 & 4 & 5 \end{pmatrix}.$$

Multiplication of a matrix by its transpose $(\mathbf{A}\mathbf{A}^T)$ is important because a **square** and **symmetric matrix** with interaction information can be obtained. To first appreciate how a square and symmetric matrix is obtained, consider this particular 2×3 matrix, $\mathbf{A}$, multiplied by the 3×2 transpose, $\mathbf{A}^T$:

$$\mathbf{A}\,\mathbf{A}^\mathbf{T} = \begin{pmatrix} 1 & 2 \\ 1 & 0 \\ 0 & 1 \end{pmatrix} \begin{pmatrix} 1 & 1 & 0 \\ 2 & 0 & 1 \end{pmatrix} = \begin{pmatrix} 5 & 1 & 2 \\ 1 & 1 & 0 \\ 2 & 0 & 1 \end{pmatrix}$$

A 3×3 square matrix is yielded by the multiplication, where each element off the 5, 1, 1 diagonal has an equal counterpart, i.e., in row 1, column 2 and row 2, column 1, the value "1" appears. Notice that the **diagonal** in a square matrix is conventionally viewed as starting from the upper left corner and ending at the lower right corner, as shown.

Now to acquire a sense for how such matrix multiplication provides information on how the elements may relate to each other, assume that the following gene expression data is obtained from an experiment in which g_{ij} represents the expression of gene-i under condition-j:

	Condition 1	Condition 2
Gene 1	g_{11}	g_{12}
Gene 2	g_{21}	g_{22}

A matrix, **A**, can describe the preceding gene expression data. The product $\mathbf{AA}^\mathbf{T}$ is

$$\mathbf{AA}^\mathbf{T} = \begin{pmatrix} g_{11} & g_{12} \\ g_{21} & g_{22} \end{pmatrix} \begin{pmatrix} g_{11} & g_{21} \\ g_{12} & g_{22} \end{pmatrix} = \begin{pmatrix} (g_{11}^2 + g_{12}^2) & (g_{11}g_{21} + g_{12}g_{22}) \\ (g_{11}g_{21} + g_{12}g_{22}) & (g_{21}^2 + g_{22}^2) \end{pmatrix}.$$

$$(8.1)$$

The diagonal element, $(g_{11}^2 + g_{12}^2)$ provides a gauge of gene-1's behavior under different conditions. The other diagonal element, $(g_{21}^2 + g_{22}^2)$, provides a gauge of gene-2's behavior under different conditions. The off-diagonal element, $(g_{11}g_{21} + g_{12}g_{22})$, provides an indication of how the expression of gene-1 and gene-2 may be related by co-expression. Therefore, obtaining $\mathbf{AA}^\mathbf{T}$ looks like a reasonable start for processing the data in order to decompose it into trends.

Note that there are different ways to process the information in the data, **A**. In the example, it was shown how $\mathbf{AA}^\mathbf{T}$ produced a square matrix where gene specific and co-expression information was present in the elements. A **correlation matrix** contains somewhat similar information, but the data are processed differently and thus, the matrix elements have different values and ranges.

A **correlation matrix** contains the values of the **correlation coefficients** between variables. Correlation coefficients quantify the relationships between observed values of variables. When a correlation coefficient equals zero, the two variables behave as if they have no relationship to each other. When one variable increases or decreases in value, the other variable exhibits no net tendency to increase or decrease. A correlation coefficient value equal to one indicates that two variables track each other perfectly; when one variable doubles in value, the other variable doubles in value as well. The value of a variable may instead tend to increase as another variable decreases, which is a negative correlation.

A correlation matrix is also square and symmetric. The diagonal elements equal 1 because a variable will always correlate with itself. For three variables, x, y, and z, and different conditions, a square correlation matrix might look like

$$\begin{pmatrix} xx & xy & xz \\ yx & yy & yz \\ zx & zy & zz \end{pmatrix} = \begin{pmatrix} 1 & 0 & 0.2 \\ 0 & 1 & 0.9 \\ 0.2 & 0.9 & 1 \end{pmatrix}.$$

The matrix indicates that x and z are weakly correlated, whereas y and z are fairly strongly correlated.

As an aside, you may now be wondering why data analysis can be challenging when pairwise correlations can be calculated. For example, two correlation matrices are given here:

$$\begin{pmatrix} xx & xy & xz \\ yx & yy & yz \\ zx & zy & zz \end{pmatrix} \qquad M_1 = \begin{pmatrix} 1 & 0 & 0 \\ 0 & 1 & 0 \\ 0 & 0 & 1 \end{pmatrix} \qquad M_2 = \begin{pmatrix} 1 & 1 & 1 \\ 1 & 1 & 1 \\ 1 & 1 & 1 \end{pmatrix}.$$

M_1 indicates that y is unrelated to x and z is unrelated to x. Therefore, y is unrelated to z. Consequently, there are three unrelated, independent "things" going on in the data. In contrast, M_2 reveals that y and x track each other and z and x correlate perfectly. Therefore, y and x track each other. In this case, there seems to be only one "thing" going on. Tracking one variable allows one to also make statements about what the other two variables are doing. That seemed easy enough, but consider a 1000×1000 matrix. There will be about 500,000 relationships between variables, and sorting through all that information in the way done for the 3×3 system would take a long time, if it could be even be done. Moreover, the examples with M_1 and M_2 were based on the extreme limits of the correlation matrices: all variables were either totally uncorrelated or all perfectly correlated. Imagine sorting information on 1000s of variables when a lot of "gray" areas exist. There has to be a better way to establish how many "things" are going on in the data and how strong one "thing" is compared to another.

Overall, AA^T is a square matrix, or data can be processed to yield a square correlation matrix. Both matrices contain useful information about gene cluster or condition-based trends. Just how the initial data are processed, and the other details that differentiate current data analysis methods (e.g., singular-value decomposition and principal component analysis), are beyond the scope of this text. However, eigenvalue analysis, which is an overarching component of many methods, can be graphically understood. The next section explains what eigen value analysis is and how it contributes, in part, to the organization and interpretation of large data sets.

Data trends can be extracted and condensed by eigenvalue analysis.
Consider the following two different correlation matrices, where there are two different conditions and two genes in experiments performed with normal cells and malignant cells:

Normal, M_N		Malignant, M_M	
1	0	1	0.5
0	1	0.5	1

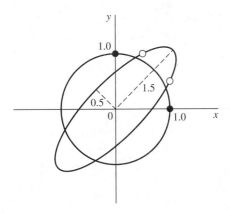

FIGURE 8.6 Ellipses passed through the points in two different correlation matrices, M_N and M_{M_0}.

A two-dimensional graphical problem can be constructed by assuming that each column contains an (x, y) coordinate. Plotting the coordinates $(1, 0)$ and $(0,1)$ for normal cells and $(1, 0.5)$ and $(0.5, 1)$ for malignant cells, and then fitting an ellipse through the points, generates Figure 8.6.

Why pass an ellipse through the points? Notice that as the data become more correlated (i.e., off-diagonal values in malignant cells > normal cells), the ellipse flattens. Geometrically, "flattening" corresponds to the one axis of the ellipse (1.5) exceeding the value of the other axis (0.5). Therefore, two dimensions are needed to describe the behavior of the variables in M_N. Here, "dimensions" means that the x-y axes can be rotated to correspond to the ellipse axes. If the axes of the ellipse are equal, then both the rotated x and y axes will be needed to specify coordinates in the shape. In contrast, M_M is tending to flatten into a line; hence, one dimension tends to cover most of the variation in the data. That is, when the x and y axes are rotated to correspond to the major (1.5) and minor axes (0.5) of the ellipse, the x axis covers much more territory in the data. Put another way, the pattern in the M_M data is simpler, which should be evident. Both genes are expressed under both conditions, so there is a tendency for both to always be somewhat or fully "ON" regardless of the conditions tested. Thus, the gene products are co-expressed and changing conditions tend to have less of an effect on the outcome. Overall, ascertaining how many major dimensions are needed to describe the variation in data will provide an indication of how uncorrelated/varied the structure is within the data.

Because a lot of data are generated from DNA arrays, it is impractical to graph and visually inspect the data for principal trends. This impracticality is surmounted, in part, by employing **eigenvalue analysis**. Recall that for a square matrix (M), an eigen value, λ, exists if a vector c exists such that

$$Mc = \lambda c. \qquad (8.2)$$

Equation (8.2) has a physical interpretation. A matrix-vector multiplication produces a new vector. The new vector can be stretched or rotated. For example, the following multiplications alters the vector $(1,1)$, which has length $2^{1/2}$ and assumes a 45-degree angle with respect to the x-y axes, by either stretching, or rotating and stretching, the original vector:

$$\underline{\text{Stretching}} \qquad\qquad \underline{\text{Stretch and change direction}}$$

$$\begin{pmatrix} 2 & 0 \\ 0 & 2 \end{pmatrix}\begin{pmatrix} 1 \\ 1 \end{pmatrix} = \begin{pmatrix} 2 \\ 2 \end{pmatrix} \qquad \begin{pmatrix} 1 & 2 \\ 3 & 4 \end{pmatrix}\begin{pmatrix} 1 \\ 1 \end{pmatrix} = \begin{pmatrix} 3 \\ 7 \end{pmatrix}$$

Therefore, Equation (8.2) says when a square matrix multiplies a particular vector, only stretching of the vector by a factor, λ, occurs. The **eigenvalues** can be computed as follows, where *det* is shorthand for computing a determinant:

$$\mathbf{Mc} = \lambda\mathbf{c} = \lambda\mathbf{I}\,\mathbf{c}. \tag{8.3a}$$

$$(\mathbf{M} - \lambda\mathbf{I})\mathbf{c} = 0. \tag{8.3b}$$

$$det\,(\mathbf{M} - \lambda\mathbf{I}) = 0. \tag{8.3c}$$

To provide an example, the eigen values for $\mathbf{M_M}$ are found as follows:

$$(\mathbf{M_M} - \lambda\mathbf{I}) = \begin{pmatrix} 1 - \lambda & 0.5 \\ 0.5 & 1 - \lambda. \end{pmatrix}$$

$$det\,(\mathbf{M_M} - \lambda\mathbf{I}) = (1 - \lambda)^2 - 0.25 = 0;\ \lambda_1 = 0.5\ \text{and}\ \lambda_2 = 1.5.$$

We find that there are two eigenvalues. In general, an $n \times n$ square matrix will have n eigenvalues. Moreover, the eigenvalues sum to 2 (i.e., n) when the diagonal elements are 1 as in the 2×2 example provided above. A more interesting observation is the two eigenvalues correspond to the lengths of the major and minor axes of the ellipse that passed through the data in $\mathbf{M_M}$. Therefore, we can compute the eigen values associated with a square and symmetrical, $n \times n$ correlation matrix. There will be n eigenvalues that sum to n and we can visualize the eigenvalues to be the lengths of the axes that the data surround.

Comparison of the relative values of the eigenvalues provides insights into the structure and relationships buried in all the data. For example, in the case of $\mathbf{M_M}$, there is a dominant eigenvalue (1.5 vs. 0.5) suggesting that the data are somewhat well-correlated and easy to describe (e.g., both genes tend to be "ON" under the conditions used). In contrast, the two eigenvalues for $\mathbf{M_N}$ are both 1, which roughly indicates that the data vary equally in two directions. That is, extensive variation of the data occurs along two independent dimensions in $\mathbf{M_N}$, whereas more variation occurs along one axis than another in $\mathbf{M_M}$.

Current directions in bioinformatics.   Bioinformatics is a rapidly moving and widening field. Some contributors are striving to develop other information gathering methods. For example, DNA arrays now provide information on the relative levels of mRNAs from different genes. However, a ten-fold higher level of one mRNA versus another may not always mean that there is a ten-fold difference in the levels of proteins encoded by the different mRNAs, because the translation of mRNAs to proteins involves many steps and competition for ribosomes. Therefore, experimentalists are striving to develop devices that provide a "picture" of the types and relative levels of the proteins actually present. The inventorying of what proteins are actually present and the analysis of such data is called **proteomics**. On the computational side, workers are busy examining the information present in sequenced genomes. Matrix algebra and statistical tools are being adapted and honed to discover insights buried within extremely large data sets.

8.6 | Future Studies

If you are interested in pursuing metabolic or tissue engineering further, a course in **biochemistry** will expose you to the full spectrum of biochemical reactions and molecules found in cells. To pursue metabolic engineering further, courses in **genetics** and **recombinant DNA methods** will prepare you for understanding how to decipher and modify a cell's genetic repertoire. A course that covers **cell** (or microbial) **physiology** will provide a start on integrating the multitude of molecular mechanisms into a coherent systems picture. Finally, an engineering or mathematics elective course that covers aspects of **network-flow analysis** will provide you with the background to quantitatively assess metabolic engineering strategies. **Linear programming (LP)** is a common topic in network analysis courses, and many bioengineers have become conversant in LP methods and software programs.

Students interested in understanding tissue engineering in more detail would benefit from studying **organic** and/or **polymer chemistry** in order to grasp further the current paradigm of using scaffolds to direct cellular proliferation. The latter courses would introduce you to how polymers are made and engineered at the molecular level to have different physical and degradation properties. A general course in **biomaterials** would accelerate your understanding of tissue engineering as well as other materials applications in bioengineering. Finally, a life science or bioengineering course that covers **receptor-ligand signaling mechanisms** and **cellular differentiation** would provide a deeper biological background for understanding current approaches as well as new approaches that tap into biological mechanisms more than the current scaffold-cell growth approach.

Fully understanding devices such as DNA chips and then contributing to the improvement or generation of alternate devices requires a deeper understanding of biomolecular binding. Chapters 5 and 6 provide a start, which can be followed up with studies on how the extent of binding depends on molecular structure and what physical processes control how fast binding occurs. Exposure to surface microprinting techniques along with microdevice design and fabrication technologies would also be beneficial. The latter refers to the methods used to put a lot of small features such as binding spots and small flow channels on a surface. Background on these subjects can often be found (or cross-listed) within chemical and electrical engineering departments. A solid background in matrix algebra and statistics is essential for fully understanding methods such as singular value decomposition.

Some introductory and advanced textbooks are cited at the end of this chapter that can be consulted to explore further the methods and fundamental tools that are used by metabolic and tissue engineering practitioners.

REFERENCES

Improved conversion of methanol to single-cell protein by *Methylophilus methylotrophus*. Windass, J.D., Worsey, M.J., Pioli, E.M., Pioli, D, Barth, P.T., Atherton, K.T., Dart, E.C., Byrom, D., Powell, K., and Senior, P.J. *Nature.* 1980. 287: 396–401.

Cell growth and by-product formation in a pyruvate kinase mutant of *E. coli*. Zhu, T., Phalakornkule, C., Koepsel, R.R., Domach, M.M., and Ataai, M.M. *Biotechnol. Prog.* 2001. 17: 624–628.

Characterization of growth and acid formation in a *Bacillus subtilis* pyruvate kinase mutant. Fry, B., Zhu, T., Koepsel, R.R., Phalakornkule, C., Domach, M.M., and Ataai, M.M. *Applied and Environmental Microbiology.* 2001. 66: 4045–4049.

Reconstruction of damaged corneas by transplantation of autologous limbal epithelial cells. Tsai, R., Li, L., and Chen, J. *New England Journal of Medicine*. 2000. 343(2): 86–93.

Successful transplantation of bioengineered tissue replacements in patients with ocular surface disease. Schwab I.R., Reyes M., and Isseroff R.R. *Cornea*. 2000. 19(4): 421–426.

A fibrin-based bioengineered ocular surface with human corneal epithelial stem cells. Han B., Schwab I.R.,

Madsen T.K., and Isseroff R.R. *Cornea*. 2002. 21(5): 505-510.

Tissue-engineered cells producing complex recombinant proteins inhibit ovarian cancer *in vivo*. Stephen, A.E., Masiakos, P.T., Segev, D.L., Vacanti, J.P., Donahoe, P.K., and MacLaughlin, D.T. *Proceedings of the National Academy of Science*. 2001. 98(6): 3214–3219.

ADDITIONAL READING

Metabolic Engineering: Principles and Methodologies. Gregory Stephanopoulos, Aristos Aristidou, and Jens Nielsen. Academic Press. ISBN: 0126662606 (November 1998).

Computational Modeling of Genetic and Biochemical Networks (Computational Molecular Biology). James M. Bower (Editor) and Hamid Bolouri (Editor). MIT Press; 1st Edition. ISBN: 0262024810 January 22, 2001.

An Introduction to Metabolic and Cellular Engineering. S. Cortassa, M. A. Aon, A. A. Iglesias, and D. Lloyd. World Scientific Publishing Co., Inc. ISBN: 9810248369 (March 1, 2002).

Metabolic Engineering (Bioprocess Technology), Vol. 24 by Sang Yup Lee (Editor) and Eleftherios T. Papoutsakis (Editor). Marcel Dekker. ISBN: 082477390X (September 1999).

Technological and Medical Implications of Metabolic Control Analysis. Athel Cornish-Bowden (Editor) and Maria Luz Cardenas (Editor). Kluwer Academic Publishers; 1st edition. ISBN: 0792361881 (April 15, 2000).

Bioprocess Engineering: Basic Concepts. Michael L. Shuler and Fikret Kargi. Prentice Hall; 2nd edition. ISBN: 0130819085 (October 31, 2001).

Methods of Tissue Engineering. Anthony Atala (Editor) and Robert P. Lanza (Editor). Academic Press; 1st edition. ISBN: 0124366368 (October 2001).

Polymer Based Systems on Tissue Engineering, Replacement and Regeneration. Rui L. Reis. Kluwer Academic Publishers. ISBN: 140201001X (February 2003).

Tissue Engineering of Vascular Prosthetic Grafts. P. P. Zilla (Editor) and Howard P. Greisler (Editor). R. G. Landes Co; 1st edition. ISBN: 1570595496 (May 15, 1999).

DNA Microarrays and Gene Expression: From Experiments to Data Analysis and Modeling. Pierre Baldi, G. Wesley Hatfield, Wesley G. Hatfield. Cambridge University Press. ISBN: 0521800226 (October 2002).

DNA Array Image Analysis: Nuts & Bolts. Gerda Kamberova and Shishir Shah. DNA Press. ISBN: 0966402758 (October 10, 2002).

A Biologist's Guide to Analysis of DNA Microarray Data. Steen Knudsen. John Wiley & Sons; 1st edition. ISBN: 0471224901 (March 22, 2002).

Bioinformatics Computing. Bryan P. Bergeron. Prentice Hall PTR; 1st edition. ISBN: 0131008250 (November 19, 2002).

EXERCISES

8.1 The yield for growth on methanol is 0.4 g cell/g methanol. If one million pounds of methanol is used per year as a feed for a microbial process that produces protein for animal feed, how many extra pounds of protein can be produced if metabolic engineering increases the cell yield by five percent? What is the extra mass of protein in human equivalents, assuming an average human contains 12.6 kg of protein?

8.2 A typical cell yield for the growth of *E. coli* on glucose is 0.3 g cell dry wt/g glucose when some glucose is lost to carbon by-products such as acetic acid (CH_3COOH; MW = 60).

(a) Assuming that the loss entails 40 percent of glucose carbon going to acetate, what fraction of glucose carbon goes to cell mass and carbon dioxide?

(b) If acetate production is stopped by metabolic engineering, and CO_2 production per gram of total glucose used remains fixed, what is the cell yield that can be attained and by what factor does it compare to the "wild-type" cell yield? Comment on energetic implications of this estimate.

8.3 What might happen to acetate production and cell yield on glucose if instead of deleting pyruvate kinase from a cell, the capacity of the TCA cycle was increased?

8.4 Why is developing a tissue engineering strategy for retina replacement potentially easier than that for engineering a replacement for a highly blood-infused organ such as a liver?

8.5 When transfected cells from a donor are implanted into a patient, note three hurdles that you anticipate must be surmounted for the technology to work.

8.6 The alcohol propanol (CH_3—CH_2—CH_2OH) and the carboxylic acid acetic acid (CH_3—COOH) combine during an esterification reaction. What is the chemical structure of the reaction product?

8.7 To follow up Figure 8.4, lactide and glycolide can be chemically reacted to form polylactide and PGA, respectively. Lactide and glycolide can also be reacted *together* to form a polymer that possesses remnants of both monomers joined by an ester linkage. Such a polymer is called a copolymer. Can you think of reasons why copolymers are synthesized for biomaterial and tissue engineering applications?

8.8 A restriction enzyme cuts after the sequence, TTAA. Show how restriction enzyme treatment will alter the following DNA, which will be done for cloning DNA in metabolic engineering:

GCATACCGTTAAGCGCTAAT
CGTATGGCAATTCGCGATTA

8.9 The yield of a microbe is increased from 0.3 to 0.6 g cell dry wt/g glucose, because energetic inefficiencies have been reduced. The metabolically engineered cell has the same kinetics (exponential/internal control rate and external control dependence on glucose) as the unengineered cell. Draw curves that correspond to the (1) unengineered and (2) engineered cells when growth occurs on glucose in a flask. Assume the initial concentrations of (1) and (2) are equal and each has been provided with the same initial amount (grams/liter) of glucose.

8.10 What are the two methods used to introduce foreign DNA into bacterial cells?

8.11 What is the transpose of the following matrix?

$$\begin{pmatrix} 1 & 2 & 1 \\ 2 & 2 & 2 \\ 3 & 3 & 3 \end{pmatrix}.$$

8.12 For the matrix answer the following questions:

$$\mathbf{A} = \begin{pmatrix} 1 & 1 \\ 0 & 1 \\ 1 & 0 \end{pmatrix}.$$

(a) What is $\mathbf{AA}^T$? Point out the diagonal and explain why it is symmetric.
(b) What are the eigenvalues, and do they sum to one?
(c) Each element in $\mathbf{AA}^T$ is reduced by the same factor to rescale the data in the matrix. The factor makes the elements on the diagonal equal to 1. What are the eigenvalues, and do they sum to the dimension of $\mathbf{AA}^T$?

8.13 The following correlation matrix $\mathbf{A}$ was generated:

$$\begin{pmatrix} 1 & 1 \\ 1 & 1 \end{pmatrix}.$$

(a) What are the eigenvalues of $\mathbf{A}$?
(b) What is the interpretation of the relative values of the eigenvalues?

8.14 The following correlation matrix was obtained:

$$\begin{pmatrix} 1 & 1 & 1 \\ 1 & 1 & 1 \\ 1 & 1 & 1 \end{pmatrix}.$$

Without doing computations, state how many eigenvalues there are and their values.

8.15 The following correlation matrix was obtained:

$$\mathbf{M} = \begin{pmatrix} 1 & 0.9 & 0.3 \\ 0.9 & 1 & 0.6 \\ 0.3 & 0.6 & 1 \end{pmatrix}.$$

The eigenvalues are 2.34, 0.73, and 0.04. Explain the significance of the relative values of the eigenvalues.

Medical Engineering

PART

IV

Primer on Organs and Function

9.1 | Purpose of This Chapter

The first section of this text began with a review of biochemistry, metabolism, and cell biology. This review enabled the understanding and pursuit of bioengineering activities such as diagnostic tool design, drug development, chemical warfare neutralization, and metabolic engineering. Now, we turn to engineering at the human scale. In this domain, replacing or supplanting the function of diseased organs, diagnostic imaging, and other efforts are pursued by bioengineers.

To understand engineering efforts at the human-scale, it is helpful to understand how cellular and molecular phenomena provide functionality within the broader organization of organs and body function. Overall, bioengineers often have to link the cellular, organ, and whole-human scales when "*something must work on the human scale*" while "*working with or harnessing, in part, what occurs on the cellular scale.*" For example, as will be discussed in Chapter 12, the materials used to replace diseased joints must possess adequate strength and other physical properties. However, how a material interacts with the molecules and cells of the immune system can have an equally important bearing on how well the implanted material will actually work.

In Chapter 1, the building blocks of cells and basic cells types were presented. In Chapter 3, biological systems were described as constant users of material and energy in order to maintain organization and functionality (i.e., combat entropy), and control over this activity must be provided. Although the human body is more complex than any one cell that it contains, a similar top-down and functional overview of the human body provides a useful organization, and will be used in this chapter.

After an inventory analysis of the human body, an overview will be provided regarding how the physical and cellular systems that constitute the digestive system provide the body with raw materials for biosynthesis and energy. How raw materials are conveyed to other cells is outlined in the section on circulatory systems. Two important control systems will then be discussed. The first involves how the volume and composition of the fluid that provides cells with nutrients and then absorbs wastes, that is, the blood, are controlled via the renal system. This chapter concludes with a description of the second control mechanism: how the ongoing activities of cells and organs are coordinated. There, protein-ligand binding reemerges as an important mechanism. Moreover, when control circuits are linked, interesting new behaviors emerge. Subsequent courses in physiology will enlarge upon the background provided in this chapter and describe further how whole-system function is linked to cellular and metabolic phenomena.

9.2 Basic Parameters and Inventories in the Human Body

In the traditional biomedical engineering literature lies the definition of the **standard man**. The parameters and other characteristics are an agreed upon set of average values for a male. These values provide a basis for calculations and design. When the case arises where the subject of interest is not average, scaling procedures, some of which are presented in the next chapter, can be used to adjust the standardized values. Values for a standard female also exist.

Table 9.1 and Figure 9.1 provide information on a standard male. From Table 9.1, the average male is 60–70 percent water by mass, which of course reflects

TABLE 9.1*

Standard Man Parameters	
Parameter	Value
Age	30 years
Height	1.73 m
Weight	150 lb, 68 kg
Surface area	19.5 ft^2, 1.8 m^2
Body core temperature	37 °C
Skin temperature	34.2 °C
Heat capacity	0.86 kcal/kg °C
Percent body fat	12% (8.2 kg)
Fat layer	5 mm
Fluid inside cells (intracellular)	28 liters
Fluid between cells (interstitial)	10 liters
Sweat, spinal, and other transcellular fluids	minor
Blood without red and other cells (plasma)	3 liters
Basal metabolism	72 kcal/h
Oxygen consumption	250 ml/min
Total blood volume (plasma + cells)	5 liters
Resting heart output	5 liters/min
Lung capacity	6000 ml
Tidal volume	500 ml
Dead volume	150 ml
Breathing frequency	12/min

*Adapted from "Biomedical Engineering Principles: An Introduction to Fluid, Heat, and Mass Transport Processes" by D.O. Cooney.

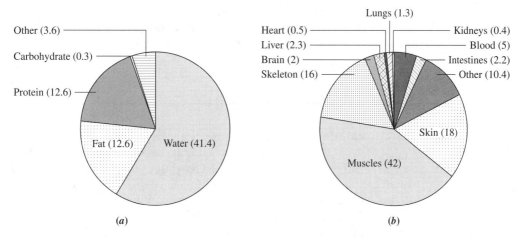

FIGURE 9.1   Breakdown of masses in the human body. (a) The relative abundance of molecular constituents; values shown are kg of molecule per 70 kg of total body weight. (b) The relative percentages of total body weight for the major organs. Values from "Biomedical Engineering Principles: An Introduction to Fluid, Heat, and Mass Transport Processes" by D.O. Cooney.

the fact that the cells that constitute the body are typically 70 percent water (Chapter 1). The heat capacity of a human (0.86 kcal/kg °C) is also close to that of water. From Table 9.1, one also finds that the blood volume (5 liters) is circulated by the heart about once per minute. Therefore, the entire volume of blood must be recharged with oxygen and nutrients every minute.

Insights are also provided on how the lungs function. During breathing, the entire lung volume of 6 liters is not evacuated and refilled. The elasticity of the lungs and airways permits the inhalation and exhalation of 500 ml per breath. This inhalation/exhalation volume is commonly referred to as the **tidal volume**. Of the 500 ml inhaled, 150 ml fills the airways leading to the lungs, which leaves 350 ml to exchange with the gases in the lungs. Therefore, the number of breathing cycles required to inhale a total volume equal to the lung volume is $6000/(500 - 150) = 17$. When the breathing rate is 12/minutes, it thus takes $17/12 = 1.4$ minutes or more to fully replace the air in the lungs.

This turnover time can have several impacts. First, carbon dioxide from cellular respiration can accumulate in the lungs. Carbon dioxide is continually released from the blood during passage through the lungs, whereas a number of breathing cycles is required to "flush out" the lungs. Thus, carbon dioxide builds up until a balance between input by respiration and output via breathing is reached. The buildup of carbon dioxide triggers the breathing reflex. Another implication is that if toxin-containing air is inhaled, pulmonary ventilation may need to be increased in order to flush the lungs and to avoid the significant transfer of toxin from the lungs to the blood stream.

The molecular breakdown of the body is shown in Figure 9.1(a). The values are masses in kg of each constituent per 70 kg of total body mass. Again, water is the most abundant molecule. Fat and protein are about equally abundant.

The percentages of total body mass for different organs and anatomical components are shown in Figure 9.1(b). Muscles, skin, and the skeleton dominate. Blood, which many forget is actually a tissue, contributes considerable mass compared to brain, liver, heart, lungs, kidneys, and intestines. The brain contributes only about two percent of the mass of the body, yet its capabilities and responsibilities are remarkably extensive.

9.3 | Digestive System

Functional description. An overview of the digestive system is shown in Figure 9.2. The digestive process actually begins when food enters the mouth. Salivary glands release enzymes called **amylases** that can degrade sugar polymers to smaller, simple sugars. Interestingly, similar enzymes, but from a different source, are used in beverage fermentation processes. When beer is made, for example, an early step in the process involves breaking down the complex, plant-derived and other sugars via the amylases that are present in the barley used, or added to accelerate the process. Once broken down, yeast can ferment the simpler sugars to produce more yeast, carbon dioxide, and alcohol.

After traveling through the **esophagus**, food enters the stomach. Other enzymes are introduced that can degrade proteins into amino acids. The enzymes are collectively referred to as **pepsin**. **Hydrochloric acid** (HCl) is also secreted to further foster the breakdown of food particles so that the molecules are more prone to

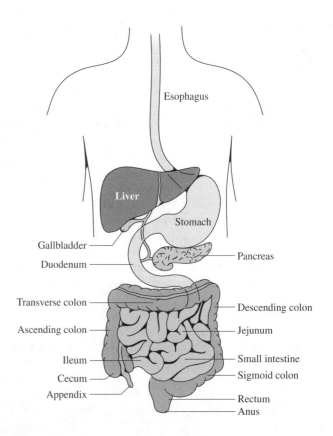

FIGURE 9.2 Gross anatomy of the digestive system.

enzymatic attack. Further digestion occurs in the first section of the small intestine, the **duodenum**. The **liver** and **pancreas** secrete a number of substances that aid digestion. These substances are carried by the fluid, **bile**, which enters the duodenum through the **common bile duct**. Bile contains bicarbonate, which neutralizes the acid added by the stomach in a manner akin to how baking soda neutralizes acid. **Bile salts** and **pancreatic lipase** are also present in bile. The former helps to solubilize fat globules much like how a detergent breaks up oil and grease droplets in water. The enzyme, pancreatic lipase, breaks down large fat molecules into smaller molecules that can be absorbed and processed further by enzyme-catalyzed reactions. In between food intake and digestion events, bile salts are stored in the **gall bladder**. The rest of the digestive track provides space, time, and surface area for nutrient absorption by the body. Water removal occurs in the large intestine, leading to the concentration of unused and indigestible material, which is then eliminated.

The length of the system in adults is 15–20 feet. About 8 liters of ingested water and saliva enter the small intestine per day. The American diet contains 200–800 g carbohydrate, 120 g protein, and 20–160 g fat per day. Thus, the mass throughput is about 9 kg/day, which is more than 10 percent of a 70-kg adult's body weight. This is comparable to the fuel to weight ratio for an automobile that burns one tank of gas per day.

Example of cellular activity that confers digestive system physical function. The **parietal cells** that line the stomach are responsible for acid secretion. About one billion cells are present and they are capable of secreting 20 mmols of hydrochloric acid per hour. Within these cells, ATP hydrolysis is used to drive H^+ from inside the cells to the stomach interior. Each cell can secrete 3 billion H^+ ions per second. The overall ion exchange and pumping is shown in Figure 9.3. Some parallels with the ion translocation processes that occur across the cell membrane that were discussed in Chapter 4 (e.g., ATPase) should be evident. Bicarbonate (HCO_3^-) and H^+ ions are generated by the activity of the enzyme **carbonic anhydrase**. As H^+ ions are pumped out, potassium ions are imported; hence, one positive ion is exchanged for another. After passage into the cell from the blood in the HCO_3^-/Cl^- exchange process, chloride ions are then re-exported to the stomach. Recall from Chapter 7 that excessive acid secretion can damage tissues and result in **erosive esophagitis**. Drugs have been designed to inhibit the proton pumps in order to lower acid secretion.

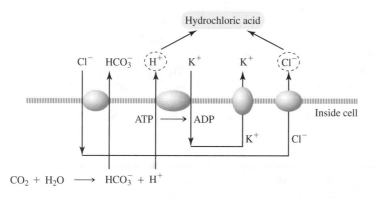

FIGURE 9.3 Ion movements across the parietal cells that are found in the stomach lining. After producing bicarbonate (HCO_3^-) via the carbonic anhydrase-catalyzed reaction, bicarbonate is exchanged for chloride. H^+ is exported via a K^+/H^+/ATPase ion "pump."

Once past the stomach, the interior lining of the small intestine has a large number of protrusions extending from it. The lugs on the bottom of a hiking boot or sneaker is a rough physical model when the number of lugs per area is increased and the size of each lug is decreased. The effect of surface texture and roughness will be encountered again in Chapter 12, where biomaterials engineering is covered. Each protrusion is called a **villus**, which is not to be confused with a vacation house. Collectively the protrusions are known as **villi**. The villi result in a surface area that exceeds by more than 500-fold the geometric area (e.g., $\pi r^2 l$ for a tubular intestine of radius r and length l). The result is that the surface area available for nutrient absorption is vastly increased. The interior lining and the multitude of protrusions are composed of **epithelial cells**.

Example of cellular activity that regulates nutrient absorption. In Chapter 2, the fate of iron in the body was used as an example for solving mass balance problems with purge and recycle. The molecular mechanisms that occur in the digestive system's cells are responsible for dictating how much of ingested iron is actually eliminated versus absorbed (i.e., M_{e1} compared to M_i in Chapter 2). Iron atoms are imported into **intestinal epithelial cells**. Within these cells, iron binds to a protein called **ferritin**, with high affinity and in a manner akin to the mechanisms described in Chapters 5 and 6.

Epithelial cells are constantly destroyed and replaced. Indeed, 17 billion cells are destroyed per day and the entire lining is replaced about weekly. When the cells are destroyed and eliminated from the body in the feces, the ferritin-bound iron exits with them. The iron that does not bind is free to enter the blood stream by migrating from the digestive system. When iron is ample, epithelial cells increase the transcription of the gene that encodes transferrin; hence, ingested iron is sequestered and eliminated. When iron is required, the opposite occurs, thereby leading to increased absorption by the body. Therefore, the material balance analysis can be used to assess what the regulation of ferritin transcription should accomplish for a given recycle efficiency and iron intake. What is not accounted for in the model in Chapter 2 is the storage of iron in the liver. Such storage provides another means to replenish the iron that is lost during recycling, and provides another example of the use of a capacitance (recall Chapter 3) in a system.

The chemical form of the iron in a food source is another factor that dictates how much ingested iron is eliminated versus absorbed. A significant fraction of the iron found in eggs, for example, is combined with phosphate. This phosphate-combined form of iron is difficult to absorb. The forms of iron found in some plants can also be difficult to absorb. In contrast, the iron present in beef liver and other protein sources is more absorbable. Thus, the mass percentage of iron or another substance in a food product can be misleading. If the content is high, but the chemical combination the substance is found in makes it difficult to extract the substance, then most of the substance will pass through the digestive system unused. The term **bioavailability** refers to the variability in usefulness of substances that are found, or can be produced, in different chemical forms. A substance with high bioavailability indicates that the majority of the main nutrient(s) found in the substance can be utilized by the body.

Once iron has entered the blood stream and organs, it is difficult to remove, which suggests why molecular control mechanisms are used to regulate input and why recycling a cautiously controlled inventory is advantageous. When iron concentration in the tissues is too high, toxicity can also result and the long-term effects can lead to organ damage and death. The accumulation of excess iron and onset of toxicity is referred to as **hemochromatosis.** Genetic mutations can lead to hemochromatosis and the disease is manageable if diagnosed early. More general information on iron and diet is available from the National Institutes of Health Clinical Center (http://www.cc.nih.gov/ccc/supplements/iron.html#provide).

9.4 | Circulatory Systems

Oxygenated blood from the heart is conveyed through vessels that branch into smaller and smaller vessels. The smallest vessels are **capillaries,** and these small vessels provide the cells within organs with oxygen and nutrients. A capillary network collects blood from organs and tissues and returns the blood to the heart-lung circulation, where reoxygenation occurs. The large supply vessels are called **arteries** and **veins** return blood to the heart.

Digested carbohydrates, fats, and proteins enter the blood stream by passing through the cells that line the small intestine into the proximal capillary networks. Unlike other organs, the **venous drainage** from the small intestine is not returned directly to the heart. Rather the exiting blood travels through the **hepatic portal vein** to the liver. One consequence of passing the blood from the digestive system through the liver is that enzyme-catalyzed reactions in the liver can contend with some potentially toxic substances that have been absorbed from the digestive system before further distribution throughout the body occurs.

However, not all the molecules derived from the digestive system are processed by the liver. Some molecules can also enter the lymphatic system. This system is the network of vessels that circulates the fluid that resides between cells. This fluid is known as the **interstitial fluid.** About 80 percent of the fluid outside cells is interstitial fluid. Some of this fluid is derived from the seepage of water through the blood vessels into the cellular space. Lymphatic capillaries bear a resemblance to blood capillaries in that the interior lining is composed of **endothelial cells.** However, lymphatic vessels are more porous and permit the infiltration and egress of large protein molecules.

The lymph circulation eventually joins the blood circulation. Two large lymphatic ducts, which have collected all the lymph flows, merge and drain into veins (subclavian) that are located in the lower neck. One-way valves permit the lymph to flow into the veins while also preventing the blood from entering the lymphatic system. The junction of the lymphatic system with the blood circulation system allows for the water and proteins that leaked out of blood capillaries into tissues to be returned back to the blood.

The lymphatic system also plays a significant role in the body's ability to fight infection. Lymphocyte division occurs in **lymph nodes,** which are organs connected by the lymph circulatory system. **Lymphocyte**s are specialized cells that have the ability to recognize the presence of a particular foreign substance. Recognition occurs when a surface protein (i.e., receptor) on a lymphocyte binds a molecule or fragment associated with an invader. After binding occurs, lymphocyte activation occurs and the immune system is mobilized. More details on the immune system will be provided in Chapter 12, which deals with biomaterials engineering.

Overall, there are two fluid flow circuits in the body: blood and lymphatic. They are not separate circuits, however; they instead merge. Together they are vital for maintaining the body's fluid balance, delivering nutrients, and combating infections. The properties of blood and how the circulatory system is designed will be explained further in Chapter 11. A physical and biological understanding is important to the bioengineers that design artificial hearts, heart-lung bypass machines, and other systems that involve blood movement.

9.5 | Heart Structure and Function

There are two common types of pumps: (1) **positive displacement** and (2) **inertial**. Positive displacement pumps rely on the tendency of common fluids to not compress when pressure is applied. A toothpaste tube illustrates the idea. When the cap is on, it is difficult to squeeze your thumb into the tube to reduce the volume, because the paste inside does not easily compress. Rather, its density remains nearly constant when low to moderate pressures are applied; hence, the attempt to squeeze a fixed mass of paste into a smaller volume is thwarted. Now, opening the cap while squeezing reduces the volume of the container and instead of compressing the paste, some paste is ejected from the tube.

The water pump in your car is an example of an inertial pump. A rotating propeller-like object (impellor) possesses momentum and continually flings the fluid

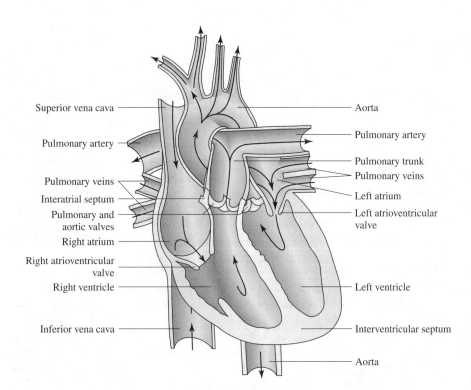

Labels (left, top to bottom): Superior vena cava, Pulmonary artery, Pulmonary veins, Interatrial septum, Pulmonary and aortic valves, Right atrium, Right atrioventricular valve, Right ventricle, Inferior vena cava

Labels (right, top to bottom): Aorta, Pulmonary artery, Pulmonary trunk, Pulmonary veins, Left atrium, Left atrioventricular valve, Left ventricle, Interventricular septum, Aorta

FIGURE 9.4 Heart anatomy and directions of blood flows in the chambers.

in contact with it out the pump. This moving fluid then travels through the cooling channels in the engine block and the radiator.

The heart is an example of a positive displacement pump. Figure 9.4 shows that the heart has four chambers; right and left **atriums** and **ventricles**. When the left ventricle contracts, blood is squeezed out into the **systemic circulation**, which services all tissues and organs except the lungs. This blood is returned to the right atrium. When the right ventricle contracts, blood is squeezed into the **pulmonary artery**, which conveys blood to a network of capillaries in the lungs. During this part of the voyage of blood, oxygen is absorbed by the red blood cells. The reoxygenated blood then returns to the heart via the pulmonary vein to the left atrium. The trip to the lungs and back to the heart is referred to as **pulmonary circulation**. Thus, the heart has two blood collection chambers (atriums) and two coordinated positive displacement pumps (ventricles).

From Table 9.1, a typical amount of blood in a human is five liters. When at rest, the heart circulates the blood volume about once every 20–60 seconds. The resting pumping rate of the human heart is about 25 times faster than the rate a fuel pump provides gasoline to an automobile engine when a car is moving at sixty miles per hour. Even at rest, the heart's pumping activity is impressive. Moreover, a human heart also lasts a lot longer than most cars and fuel pumps. Indeed, an average heart beats about 100,000 times per day and around 35 million times in a year.

9.6 Removal versus Preservation of Substances in the Blood

The digestive system eliminates some ingested water from the body, and the lymphatic and blood circulation systems connect, which tends to preserve fluid inventory. The **renal system**, however, plays the major role in adjusting the value of the body's water inventory. Additionally, because ions are essential for energizing cellular membranes (e.g., Figure 9.3), and they are also involved in other processes, another important function executed by the renal system is the regulation of ion concentrations in the blood.

The renal system also eliminates the excess nitrogen that accumulates in the blood. Nitrogen in the form of ammonia can be recycled within cells via the enzymatic reaction described in Chapter 8, which humans and some microbes possess:

$$NH_4^+ + \alpha\text{-ketoglutarate} + NAD(P)H \rightarrow glutamate + NAD(P)^+.$$

However, when excess nitrogen in the form of ammonia accumulates, mammals convert the built-up ammonia to urea (NH_2CONH_2), which is removed from the blood and excreted into the urine by the major organs of the renal system, the **kidneys**.

Animals that excrete urea are called **ureotelic**. Those that instead excrete ammonia, such as some fishes, are termed **ammonotelic**. Birds and reptiles excrete uric acid and they are thus called **uricotelic** organisms. Because some fossil evidence suggests that birds descended from reptiles, it is interesting that they exhibit similar waste processing biochemistry. Amphibians not only move from water to land as they develop from eggs into adults, their waste-processing biochemistry also changes. When frogs are in the tadpole stage, ammonia is excreted. As they develop further, they attain urea-excreting ability.

The cells in the kidney organize into capsular and tubular structures that permit the filtration of the cell-free portion of blood, the **plasma**. These structures are called **nephrons**. A substance can enter the urine flow through two processes, as shown in Figure 9.5. **Glomerular filtration** allows for cell- and protein-free blood constituents to enter a tubule. Within the tubule, protein-free plasma components undergo further sorting with respect to the individual substances. Some valuable substances such as water can leave the tubule and reenter the blood to avoid, for example, dehydration. Some substances can also be specifically imported from the blood across the tubule wall to further increase their abundance in the urine.

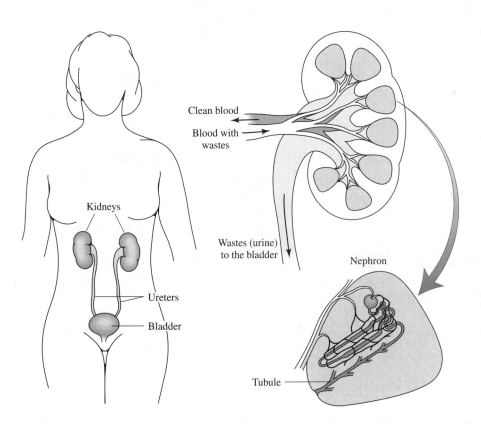

FIGURE 9.5 Kidney location (left) and structure and function (right). Blood flow to the kidney is subdivided and directed to nearly one million nephrons. Upon entering a nephron, blood first enters a sac-like structure called a Bowman's capsule. Within the capsule, physical filtration (*glomerular filtration*) occurs, leading to some cell-free fluid exiting the capsule and entering a tubule. The filtrate flowing through a tubule is in contact with capillaries (*peritubular capillary*). Constituents in the filtered fluid can enter the capillaries (*tubular reabsorption*) or be transferred from the capillary to the tubule (*tubular secretion*).

Applying the mass balance methods presented in Chapter 2, the kidneys alter the total amount of each plasma solute according to

$$\text{Net excreted} = \text{amount filtered} + \text{amount secreted} - \text{amount absorbed}. \quad (9.1)$$

Overall, the amount a given substance is excreted depends on the chemical nature of the substance and its abundance within the body. Regulatory mechanisms allow for waste excretion while retaining needed water, salts, nutrients, and blood proteins. Moreover, the retention is adaptive whereby, for example, water can be retained during periods of scarcity or eliminated when a surplus exists.

The volume of plasma cleansed of a substance per unit time is called the **renal clearance** of that substance. To illustrate further, **inulin** is a complex sugar that is not normally found in the body. Inulin is not reabsorbed or metabolized; hence, glomerular filtration is the only removal mechanism. When continuously injected into the blood stream, inulin will eventually appear in urine. The faster inulin appears in the urine and the higher the concentration, the faster the glomerular filtration rate. A mass balance on inulin can be performed to infer the glomerular filtration rate based on C_P, C_U, V_U, and Δt, which refer to inulin plasma concentration, inulin urine concentration, urine volume, and urine collection time, respectively. For example, when $C_P = 0.1$ mg/l, $C_U = 10$ mg/l, $V_U = 0.075$ liter, and $\Delta t = 1$ hour, the inulin clearance (CL_I) rate is

$$CL_I C_P = V_U C_U / \Delta t. \quad (9.2a)$$
$$CL_I = 0.075 * 10/0.1 = 7.5 \text{ l/h or } 125 \text{ ml/min.} \quad (9.2b)$$

As noted earlier, the inulin clearance corresponds to the glomerular filtration rate. This baseline measurement of inulin clearance can provide insights on how other substances are processed by the renal system. For example, if a particular substance has a clearance that exceeds the inulin clearance rate, then unlike inulin, the substance must be actively exported from the blood across the tubules into the urine. Another way to understand this is to refer to Equation 9.2a. If substance A and inulin have the same value of C_P, yet the C_U of substance A is greater than inulin, then A must have had some "help" with accumulating to a higher concentration in the urine. If substance A's clearance is less than inulin's, then some reabsorption and/or metabolism must occur. Alternately, the substance may be partially "stuck" to the blood proteins via binding. Overall, inulin clearance is established as a research tool, and clearance tests with other substances are used to diagnosis the function of the renal system.

The treatment of patients with kidney disease by dialysis is a major biomedical engineering achievement. The inventor of the artificial kidney dialysis machine, **Dr. Willem J. Kolff**, was honored in 1985 by being inducted in the National Inventors Hall of Fame (http://www.invent.org/hall_of_fame/88.html). His career began in his birthplace, the Netherlands, and then took him to the United States. The machines based on his ideas attempt to mimic the filtration of blood that occurs in normal kidneys. The concept is shown in Figure 9.6. Special membranes are used that permit the passage of small molecular weight substances such as urea into a wash solution while allowing cells and proteins to be retained in the blood. The patient's blood is pumped from the body and passed over a membrane surface. Transit over the membrane surface results in cleansing the blood prior to its return to the body.

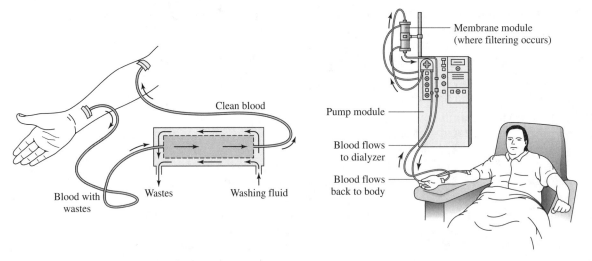

FIGURE 9.6 Operational principle of an artificial kidney and how hemodialysis is performed. (Left) The basic concept is that blood is contacted with a membrane that allows for the movement of small molecular weight substances into a wash solution while retaining cells and large proteins in the blood. (Right) A schematic of a typical kidney dialysis machine and how a patient is treated.

In currently used dialysis machines, the physical filtration functions of the kidney are reproduced, but the metabolic functions of kidney cells are not. This gap in providing full kidney function motivates tissue engineers and artificial organ designers to further improve the capabilities of human-made kidneys.

9.7 | Activity Coordination: Endocrine System

The raw materials delivered by the digestive system and circulated by blood and lymph provide the cells within tissues with raw materials and energy. Within the cells, energy transactions occur that involve ATP, as described in Chapters 3 and 4. The pathways are, in turn, regulated by binding-mediated, feedback processes akin to those shown in Figure 3.4. However, another level of control is needed to control cell replication in one organ versus another, increase or decrease cellular activity in a particular organ to cope with an environmental stimuli, coordinate muscle movement, etc. Thus, in addition to local control, the **nervous** and **endocrine** systems manage all the cellular communities to insure that coordination, development, and appropriate adjustments occur.

Here, the endocrine system will be summarized. **Endocrine glands** secrete molecules that can interact with specific cells located elsewhere in the body. The secreted molecules, **hormones**, fall into three categories based on their chemical structure: **amines**, **peptides**, and **steroids**.

The **amine hormones** are all built from the amino acid, tyrosine. Recall from Chapter 1 that tyrosine is a nonessential amino acid. Thus, the body synthesizes tyrosine, and one of its fates is to serve as a building block for hormones. **Epinephrine** and **dopamine** are representative amine hormones.

Most hormones are peptides (small proteins) or proteins. Cells often first synthesize an inactive version of the protein. The first draft is called a **preprohormone**. The first draft is converted to a **prohormone** by bond cleavage. Thereafter, additional modifications that yield an active hormone and insertion into a delivery package called a **secretory vesicle** occur. A secretory vesicle is a small spherical object that is built of cell membrane material; the interior contains proteins and other molecules that can be exported from the cell. The activation process, which cuts some bonds on the prohormone, can also produce protein fragments. Sometimes the protein fragments, which are also inserted into the secretory vesicle, have hormonal or other activities. Thus, endocrine cells are poised and ready to release a number of hormones through the expulsion of secretory vesicles. There is no delay in sending signals associated with hormone synthesis and packaging.

The steroid hormones are constructed from **cholesterol**. Thus, while high cholesterol is linked to heart disease, a serious cholesterol deficiency could also be a problem. Steriod hormones possess carbon ring chemical structures. Steroids are highly soluble in fats and lipids and insoluble in water. Due to their high solubility in fats and lipids, steroid hormones can easily cross cellular membranes; hence, secretory vesicles are not used. However, because of their low solubility in water, buildup and circulation in the blood are enabled by the binding of steroid hormones to plasma proteins. Drugs can also bind to blood proteins and affect the rate at which they are eliminated from the body, as well as the effective dosage established for target cells and tissues.

All hormones circulate in the blood. A given hormone can affect a target cell because that cell possesses a protein that can specifically bind that hormone. If a peptide hormone is involved, the receptor protein is typically on the cell surface. If the hormone is instead a steroid, the lipid-soluble hormone first passes through the cell membrane and then binds to an intracellular protein. The hormone-protein binding event initiates series of subsequent binding and chemical reactions processes that ultimately affect the target cell's activity and/or function.

The discussion so far may have prompted some recollection of ON/OFF versus continuous variable systems that were discussed in Chapter 3. Additional control features emerge for hormones and their effects. A target cell may possess more than one hormonal receptor. In some cases, the presence of hormone X will cause the receptors for another hormone (Y) to dwindle, thereby making the cell unresponsive to the presence of hormone Y. In this case, hormone X is **antagonistic** towards hormone Y. The opposite case is when some level of hormone X is required for a cell to respond to hormone Y. This is a case of **permissiveness**. To use organizational behavior analogies, some control systems are ruled by "executive veto." Others instead operate on the quorum system. Such variety introduces many interesting control logics such as double checking before an event is initiated, time-sequence development, and cycling behavior.

REFERENCES

Biomedical Engineering Principles: An Introduction to Fluid, Heat, and Mass Transport Processes. 1976, Marcel Dekker Inc, New York, NY, ISBN 0-8247-6347-5.

EXERCISES

9.1 Air containing a toxic substance is suddenly inhaled and results in the concentration in the lungs equaling 10 percent volume. How many breaths are required to reduce the toxic substance concentration to less than 1 percent volume and what is the maximum time needed?

9.2 Water, fat, and protein are dominant molecular constituents in the human body. If fat and protein are about 50 percent by weight carbon, estimate the fraction of carbon in the water-free mass of the human body and compare this value to a typical cell.

9.3 One type of an artificial heart and battery power supply weighs 4 lbs (1.8 kg). Assume that the mass of a real heart is proportional to body weight. How does the total mass of the artificial system compare to the mass of a real heart in a 70 kg adult and a 20 kg child? What are the quality of life implications for the adult and child after implantation?

9.4 The iron in food is typically broken down into two categories: (i) heme and (ii) nonheme. The heme form is akin to the form in which iron is present in your blood, and it is the most absorbable form. 100 g of tofu contains 3 mg of nonheme iron and 100 g of beef liver contains 18 mg of heme iron (British Columbia Ministry of Health File 68d, http://www. healthplanning.gov.bc.ca/hlthfile/hfile68d.html).

(a) Approximately how much tofu or liver would you have to eat daily to meet the FDA recommended value of iron intake (8 mg/day)?

(b) What possible problems could arise with a diet based exclusively on tofu or liver?

9.5 Blood flow to the kidneys is typically 1,200 ml/min. Urine flows and glomerular filtration rates are 1.5 ml/min and 125 ml/min, respectively.

(a) What percentage of the blood entering the kidneys is filtered every minute?

(b) What fraction of the total blood flow is filtered every minute?

9.6 About how long does it take to filter all the blood in the body if disease causes the glomerular filtration rate to decrease by a factor of two? What would be an adverse consequence of the lengthened filtration time?

9.7 A drug can bind to blood proteins. The fraction bound is designated by f.

(a) How will the renal clearance of the drug compare to the inulin clearance?

(b) If the glomerular filtration rate is denoted by GFR, how will the renal clearance of the drug (C_D) depend on GFR if the drug is not reabsorbed or subject to tubular secretion?

9.8 The renal clearances of four drugs were investigated in clinical trials. Some binding studies have also been performed to determine if drugs bind to plasma proteins. Some or none of these drugs may be subject to reabsorption or tubular secretion; hence, it was hoped that the clinical studies would shed more light on the drug elimination mechanisms. Data to date indicate that all of the drugs are not metabolized in the body. Based on the data in the table below, indicate what mechanisms may operate in eliminating the drugs from the body by the renal system.

Drug	Bind to Plasma Proteins	Clearance
A	NO	125 ml/min
B	NO	150 ml/min
C	YES	80 ml/min
D	YES	125 ml/min

9.9 What is the key difference between how protein/peptide and steroidal hormones are produced and presented to the body?

9.10 In one scenario, some class members vote on when a term paper is due. Based on the due date, other class members suggest how many pages it should comprise. A second scenario entails the course instructor quelling the discussion and herself setting the date and expected word count. Which scenarios are analogous to antagonistic and permissive control systems?

Biomechanics

CHAPTER

10

10.1 | Purpose of This Chapter

Humans possess moving parts that perform mechanical work. These parts are also subjected to gravitational and other forces. For example, collective limb motion and muscle action provide an athlete with the ability to temporarily oppose gravity when jumping over a high bar. In the realm that lies beyond molecules and cells, some biomedical engineers design devices and appliances that, for example, improve the functionality of people with mobility-curtailing injuries or diseases. Other biomedical engineers may design equipment that improves an athlete's performance or safety. Still others strive to perfect the artificial heart that can replace the natural heart's ability to pump blood. When biomedical engineers solve problems that lie beyond the molecular or cellular size scales, biological and physiological knowledge is kept in mind when using the principles of Newtonian mechanics and mathematical analysis tools. Some helpful places to scope the full range of biomechanical engineering activity are the *American Society of Biomechanics*, *American Society of Mechanical Engineers*, *International Society of Biomechanics*, and *Rehabilitation Engineering and Assistive Technology Society of North America*.

In this chapter, the flavor and utility of biomechanical analysis will be illustrated by examining a seemingly simple process: human walking. Concerning flavor, you will see how some key mathematical tools learned in prior courses can be applied. The analysis of walking will yield a mathematical result that is amenable to **optimization**. Indeed, engineers often mathematically model problems to ascertain the interaction between variables. One interesting case that can arise is when a variable combination exists that either minimizes or maximizes an objective. In this case, subjecting the general result to optimization analysis suggests that for a given walking speed, an individual has a particular stride length that *minimizes* (i.e., optimizes) their use of energy. Regarding utility, the result can be **scaled** to enable the comparison of two people walking. Scaling is often done to generalize results to facilitate comparisons and to answer questions such as "What happens when some factor is doubled?" In this case, an **ergonomic** problem will be posed and solved using scaling. Additionally, the result can be used as an input into another problem: designing a power supply for a mobile recipient of an artificial heart. Using the solution of a subproblem as an input into a larger problem is often called **patching** or **synthesis**.

While at the outset, the analysis of walking may look mostly like a math and physics problem, there are numerous links to biological knowledge. For example, the optimization result suggests that at any walking speed, an individual's body and brain can "compute and control" with the result that the least amount of work is performed. Does biological mechanism knowledge concur with this result or does the optimization result provide a theory-inspired basis for analyzing mechanistic information in a different way? How can this conclusion be tested in a physiologically relevant manner? The second question is addressed in the chapter by reproducing some data from the literature.

10.2 | Power Expenditure in Walking*

Observe someone walking. One leg is swung forward and the torso rises. At the end of the swing, the foot makes ground contact as the torso drops. The friction between the foot and the ground results in a short pause where the leg swing stops and the body is propelled forward. The motion can thus be broken down into two parts: (1) the leg swings with a regular frequency and (2) the torso bobs up and down against gravity. Because energy and power are scalars, the two motion components can be added arithmetically to yield the total power expenditure.

A simple representation of the dimensions and displacement associated with walking is shown in Figure 10.1, where the person's mass, walking speed, leg length, and stride length are M, v, L, and a, respectively. The angle the extended leg makes with a line perpendicular to the walk surface and how much the torso bobs up and down are γ and H, respectively. Recall that power is the rate of doing work. Here, work is done to endow the limb with kinetic energy. This energy is imparted to limb motion and transferred to horizontal displacement every a/v units of time. So the power expenditure for the limb swinging component (P_1) is

$$P_1 = \alpha M v^2/(a/v) = \alpha M v^3/a. \tag{10.1}$$

In this equation, α is a constant less than one. We insert a constant because in this problem the most convenient mass basis is that of the whole person, whereas this motion component involves only the leg. We therefore are assuming that more massive people have more massive legs.

Now the power expended to move the torso up and down (P_2) can be determined by using trigonometry. Again, power can be represented as the work done (against gravity) every a/v time units. The result is

$$P_2 = MgH/(a/v). \tag{10.2a}$$
$$H = L(1 - \cos \gamma). \tag{10.2b}$$

---

*This analysis of walking motion is based, in part, on Belleman's contribution to *The Physics of Sports* (see references at chapter end).

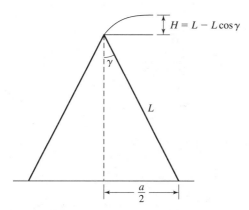

FIGURE 10.1   Simplified diagram of the displacements that accompany limb movement and torso motion during walking. The leg of length, L, swings back and forth. The step size and angle associated with the swinging are a and γ, respectively. The torso bobs up and down with a displacement, H. Adapted from A. Bellemans' analysis (*Power Demand in Walking and Pace Optimization*).

The previous two equations can be combined to yield a perfectly good result. However, from practical experience we know that the angle a leg rotates when walking is not huge. Thus, calculations would be easier to perform, and how the different variables affect the solution could be more readily determined, if the cosine function could be simplified by exploiting the engineering judgment that for most applications, γ is small. A **Taylor series expansion** followed by converting degrees into radians can provide the simplification, as shown in Figure 10.1 and the following inset:

Taylor Series Simplification.

The general Taylor series is

$$f(x) = f(x_o) + f'(x_o)[x - x_o] + \frac{1}{2}f''(x_o)[x - x_o]^2 + \ldots,$$

where x_o is the point of reference that we expand the function $f(x)$ around. In this case, Equation (10.2b) for H will be expanded about $\gamma_o = 0$, because we assume that the angle is generally small and will not depart too far from the value zero. We have

$$H(\gamma_o = 0) = L(1 - \cos 0) = 0$$
$$H'(\gamma) = L \sin \gamma$$
$$H'(\gamma_o = 0) = 0$$
$$H''(\gamma) = L \cos \gamma$$
$$H''(\gamma_o = 0) = L.$$

Putting these terms back together yields

$$H(\gamma) \approx 0 + 0[\gamma - 0] + \frac{1}{2}L[\gamma - 0]^2.$$

$$H(\gamma) \approx \frac{1}{2}L\gamma^2.$$

The above equation for H can be simplified even further by using radians for the angle γ. Recall that there are 2π radians in a circle of 360 degrees. Thus, an angle in radians equals the arc length enclosed by the angle divided by the circle radius; hence, $\gamma(\text{radians}) \approx a/2L$. Therefore, $H \approx a^2/8L = \beta a^2/L\alpha$. Note that using the Taylor series reduced a cumbersome trigonometric expression into a simpler algebraic expression where key variables attain some power, and thus their influence on the magnitude of power expenditure is more easily pictured by inspection. The price paid for using a Taylor series approximation is that as the variable exceeds the assumed baseline value ($\gamma_0 = 0$ in the case), the error increases unless we add more terms from the expansion. Generally, if we do not push the use of a simplified result too far, the advantages of simplification outweigh the losses in accuracy.

Using the simplification, the power expenditure for torso movement becomes

$$P_2 = (\beta Mga^2/L)/(a/v). \qquad (10.3)$$

Combining Equations 10.1 and 10.3 to obtain the total power expenditure results in

$$P_T = \alpha Mv^3/a + \beta Mgva/L. \qquad (10.4)$$

Note that the 1/8 factor has been lumped into another constant, β. In practice, values for α and β equal to 0.1 and 0.05, respectively, have been found to capture the magnitude of human power consumption.

10.3 Optimization Illustration: Least Power Expenditure Stride Length

Now that we have the result for total power expenditure, let's put it to work. First notice how P_T in Equation 10.4 depends on stride length, a. It appears in the denominator of the first term and the numerator of the second term. Thus, when a is small or large, one term becomes very large, resulting in a large value of total power expenditure. This outcome should make sense. Imagine walking a mile taking one inch steps or very long ones—you would tire very quickly.

The limit analysis we just performed indicates that a value of a (call it a^*) exists that would minimize the total power expenditure. The value can be found by equating the first derivative of P_T to zero, and then finding the value of a:

$$dP_T/da = 0. \tag{10.5a}$$
$$a^* = (\alpha/\beta)^{1/2}(L/g)^{1/2}v. \tag{10.5b}$$

Therefore, when the optimal stride length is used (a^*), the **minimum** power expenditure is

$$P_T{}^* = 2(\alpha\beta)^{1/2}(g/L)^{1/2}Mv^2. \tag{10.6}$$

Interestingly, your brain and body work together to actually minimize power expenditure. Another way to put this is that you have a natural gait that maximizes your use of resources, and it is a function of your mass and dimensions. This optimality has been observed in, for example, treadmill tests. In such studies, the walking speed was altered by running the tread faster or slower. How the subject shifted stride length and their power expenditure (e.g., respiration rate) were measured. Experimental data from treadmill tests, which are reproduced in Figure 10.2, clearly show that an optimum stride length exists. Another interesting outcome of the optimization analysis is that some indication is provided of how one may adjust to changes in walking speed. Equation 10.5b states that to keep power consumption minimized when pace (i.e., v) is increased by some factor, the stride length should also be increased by the same factor. However, the new value of minimum power consumption will increase by more than the factor that the pace increased, because $P_T{}^*$ depends on v^2.

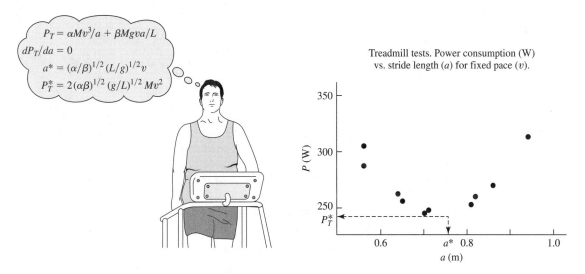

FIGURE 10.2 Treadmill data for a person walking at a pace equal to 4 km/h. Each person has a clear minimum in power consumption (w, watts) with respect to stride length, (a, meters) which was predicted by the simple biomechanical model in the text. (Adapted from Cotes and Meade, *Ergonomics* (1960): 3, 97).

10.4   Scaling the Result in an Ergonomic Analysis

The prior analysis resulted in the identification of an optimum stride length that depends on pace and a person's dimensions. Other uses of analysis results emerge readily from scaling the result. **Scaling** entails rewriting the result such that the variables are combined in important ratios that are dimensionless. As you will see, when a result is transformed in such a way, two benefits ensue. First, it is easier to determine how different situations will either correspond or differ, and if they differ, by what factor. Secondly, only *ratios* of variables are important, so a mixed bag of units can be used. That is, masses can be in pounds, lengths in centimeters, and so on.

The utility of scaling is demonstrated by an ergonomic problem. **Ergonomics** is the study and management of how people interact with the built and work environment. Since we are all built differently, when we interact with stairs, machines, etc., our bodies will experience different degrees of help or stress. Bioengineers and others deal with maximizing the benefits and minimizing the stress by accounting for human factors in the design of the machines and appliances we use.

The problem involves two different people that are required to walk the same distance in the same amount of time. Asking both people to do the same task the same way may cause greater exertion to one of the people. Thus, the question arises *"Which person expends the most energy and thus will experience more fatigue?"* Apart from one person working more than the other, such a question has practical importance because the more fatigued person may be more prone to an accident. The information from such an analysis could be used to design a rest schedule or alter the workflow such that one person is not overly physically taxed.

We can contrast the work the two people perform to address which person is most challenged by the task. One simplifying assumption that can be made for a first-pass analysis is that the values of α and β are constant and equal for both people. This assumption is not too bad for β. The model suggests β is a constant; the value was decreased to account for factors not in the model that may actually lessen power consumption, such as elastic storage of energy in the limbs. The value of α could vary with a person's mass and dimensions if, for example, someone was long or short legged. Analyses often start by making simplifying assumptions, and then additional details are added for refinement if the initial answers suggest that more work is merited. Because the product of power and time is work, and the times are the same, another simplification is that the power difference will be proportional to the work difference. Thus, we can calculate the power difference and say it is proportional to the work difference; the proportional relationship is denoted with the symbol $\sim$.

When the two people are distinguished as 1 and 2, the power/work difference when both walk at the same speed is

$$\Delta W \sim \alpha M_1 v^3/a_1 + \beta M_1 g v a_1/L_1 - \alpha M_2 v^3/a_2 - \beta M_2 g v a_2/L_2. \tag{10.7}$$

Rearranging Equation (10.7) results in

$$\Delta W \sim [\alpha v^3 M_1/a_1](1 - M_r/a_r) + [\beta g v M_1 a_1/L_1](1 - M_r a_r/L_r), \tag{10.8a}$$

where

$$M_r = M_2/M_1, \qquad \text{(10.8b)}$$
$$a_r = a_2/a_1, \text{ and} \qquad \text{(10.8c)}$$
$$L_r = L_2/L_1. \qquad \text{(10.8d)}$$

Note that the terms in parentheses do not have units; only the mass and other ratios appear. Thus, if person-2 weighs 20 percent more than person-1, then $M_r = 1.2$.

Performing the extra effort to organize results such as Equation 10.8 provides the benefits of (1) greatly simplifying calculations and (2) increasing the ease of observing variable interactions. For example, if person-2 has a mass and stride length that are both 20 percent greater than person-1, the contribution to ΔW in the first (left) term is zero; we simply do not have to worry about that part of the equation. This rapid simplification has been found *without* going through the work of plugging in numbers for a_1, a_2, v, M_1, M_2, etc. From another perspective, which will be explored soon, the first step has also been taken to answering under what circumstances will both peoples' exertion be the same.

The variable combinations that multiply the terms in the parentheses have the units of power. Often one term is used to make the whole equation dimensionless. This can be done by dividing both sides of Equation 10.8a by, for example, $\Delta t \, \alpha v^3 M_1/a_1$. That step would make all the terms in parentheses and the coefficients outside the parentheses dimensionless. The inclusion of work-time Δt also changes the proportionality into an equality. We will forego this step and just point out that whether or not this next step is taken, the ratio terms that multiple the quantities in the parentheses can be regarded as **weights**. That is, when the values of individual variables are inputted, the values of the coefficients typically exhibit a range; smaller values may result in a particular term having less importance than another.

We are now in a position to answer the original question. One way to answer the question is to uncover when both people's efforts are equal. Any other case will result in one person working more than the other. Such a screening is useful because simple rules for when a difference will occur can often be found. When a difference is indicated, then the effort of a more detailed calculation becomes justified.

A simple answer is, when the quantities in both terms bounded by the parentheses are zero. From the left and right terms, we thus obtain

$$M_r/a_r = 1. \qquad \text{(10.9a)}$$
$$M_r a_r/L_r = 1. \qquad \text{(10.9b)}$$

The above two results can be combined to obtain the following rule:

If $M_r = a_r = L_r^{1/2}$, then the two people do the same amount of work.

Is this the only possibility for the two people to exert themselves equally? No, there is the trivial solution of $M_r = a_r = L_r = 1$. The solution is trivial because it corresponds to two identically proportioned people, so there of course should be a zero difference. Indeed, it is always a good idea to check to see that an analysis gives the correct result for a trivial solution. If it does not provide an answer consistent with the problem logic at some well-defined boundary for the variables, then the odds are high that a mistake has been made in the analysis. The other nontrivial case is when the left and right terms have different signs and they sum to zero. Obtaining the result is left to you as an exercise. Note that it is consistent with the prior rule when optimal stride lengths are adopted by each worker.

From this example, you have seen that spending some time to organize the result by scaling makes it easier to ferret out the consequences of different values of variables. Whether the subject is bioengineering or airplane design, all engineers employ clever means to make the results of their analyses easier to comprehend and more succinct. In the process, design rules or decision metrics are developed.

10.5   Using the Solution to Solve a Larger Problem

Imagine that you are a member of a team whose job is to design a total replacement artificial heart. As will be described in more detail in Chapter 11, such a device is a well-engineered blood pump that can be implanted in a patient. The vision is that with the implant, the patient will be mobile and thus enjoy some appreciable quality of life as opposed to being totally bed-bound. Your assignment is to design the battery power supply; your system will power the heart so that blood circulates through the lungs and other organs, which in turn allows for oxygen transport into tissues and removal of respiration waste products such as carbon dioxide.

This is an exciting yet somewhat daunting task, because your decisions will directly affect a fellow human being who is already in a precarious situation. Where shall you start? A reasonable starting question is "How big should the battery be?" This question is reasonable, because if it is the size of a refrigerator based on what you deem the use and demand profiles to be, then it is questionable whether the quality of life objective can be attained. The question can be answered, in part, by using the prior analysis of the power expended in walking within an analysis of wider scope.

To begin, a system diagram is constructed to show the key processes and the relationship between them. Figure 10.3, which depicts an energy transfer chain, shows that the electrical energy supplied by the battery is first converted to mechanical work. That is, the battery provides the power to move the parts of the blood pump much like a battery provides the power to move the motor in a cooling fan. Then the blood pump moves the blood through the arteries at a rate that allows for the metabolic and tissue oxygenation demands of walking to be met. We assume that this demand is proportional to the power expenditure of walking, which was the subject of the first part of this chapter.

The diagram also shows that two conversion efficiencies can be envisioned. First, 100 percent of the battery's stored energy will not be converted to mechanical

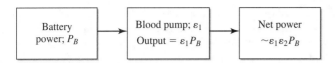

FIGURE 10.3 Simple series power transmission chain where each element has an efficiency, ε, less than 1.0.

work due to friction in the moving parts and resistance in the electrical components; hence, ε_1 is less than or equal to one. Secondly, the mechanical energy imparted by the pump for the purpose of moving blood may not be fully harnessed for walking. Doubling blood flow rate may, for example, allow for the walking rate to increase by a factor of 1.8. To account for the correspondence between blood flow rate and activity rate, the second factor, ε_2, is defined. The greater ε_2 is, the more additional work can be done when the blood flow rate increases by a particular increment. Thus, ε_1 is intrinsic to the device and ε_2 is a characteristic of the device's recipient.

The target of the analysis is to specify what battery lifetime is required. Lifetime is usually stated as the product of current and time (e.g., Ampere-hours). Mathematically, this is equal to the area under a current versus time discharge curve when a power-consuming load is present. Another way to view this definition is to visualize a portable CD player. Ten 9-volt batteries connected in parallel will supply 9 volts DC for a CD player for a longer period of time than just one battery. Thus, the trade-off for longer electrical supply lifetime for any device, including an artificial blood pump, is that a larger battery size and/or mass is needed.

Recalling that the product of current (I) and voltage (V) equals electrical power dissipation, the effective blood pumping power output (i.e., what the implant accomplishes) is

$$IV\varepsilon_1 = \text{Effective Blood Pumping Power Output.} \qquad (10.10)$$

When multiplied by ε_2, the effective amount of metabolic work (i.e., walking) that can be performed is

$$IV\,\varepsilon_1\varepsilon_2 \sim P_w. \qquad (10.11)$$

The length of time we consider the patient to be mobile with the battery system (i.e., away from a permanent power supply) is t_m. Thus, the required battery lifetime (B_L), which is the product of I and t_m (e.g., Amp-hours) is given by

$$B_L \sim P_w t_m / (V\varepsilon_1\varepsilon_2). \qquad (10.12)$$

Note that we "patched in" our prior analysis; P_w appears as a term and we have a general result for it in Equation 10.4, or its minimum value is provided by Equation (10.6). We have thus patched one problem solution to another.

Equation 10.12 has some interesting facets. First, a greater demand in terms of walking power expenditure, and/or the time over which the battery power supply is required, increases the required battery lifetime and therefore the size. The two efficiencies also multiply to result in an overall efficiency. Many cases arise where an overall efficiency proves to be the product of those associated with the underlying subsystems. One classic example is a multi-level food chain. The primary producers harness sunlight by photosynthesis. They, in turn, get consumed by their predators, and the resources are passed on through the food chain. The more levels in the chain, the fewer ultimate consumers can exist at the top because the conversion of mass and energy into new material at each level is not 100 percent efficient.

You can use some simple calculations to see how inefficiency accrues. Three levels with 80 percent efficiency per level have an overall efficiency of about 50 percent (0.80^3), which is about as efficient as six levels with 90 percent efficiency per level. Ninety percent may sound quite efficient compared to 80 percent, but subsystems connected in series can quickly become an inefficient overall system as the number of subsystems grows. The degeneration of efficiency is what drives engineering work on component improvement. Even small improvements in subsystem efficiencies can yield considerable gains in the overall efficiency and performance of a multicomponent system.

ADDITIONAL READING

The Physics of Sports. Angelo Armenti, Jr. (Editor). Springer Velag. ISBN: 0-88318-946-1 (1992).

Scaling in Biology. James H. Brown (Editor) and Geoffrey B. West (Editor). Oxford University Press. ISBN: 0195131428 (April 2000).

Mathematical Ideas in Biology. Maynard Smith. Cambridge University Press. ASIN: 0521095506 (November 1968).

Life at the Extremes: The Science of Survival. Frances Ashcroft. University of California Press. ISBN: 0-520-22234-2 (2000).

Assistive Technology: Essential Human Factors. Thomas W. King. Allyn & Bacon; 1st edition. ISBN: 0205273262 (September 14, 1998).

Occupational Ergonomics: Engineering and Administrative Controls (Principles and Applications in Engineering). Waldemar Karwowski (Editor) and William S. Marras (Editor). CRC Press. ISBN: 0849318009 (March 2003).

Principles of Animal Locomotion. R. McNeill Alexander. Princeton Univ Press. ISBN: 0691086788 (January 2003).

Basic Orthopaedic Biomechanics. Van C. Mow (Editor) and Wilson C. Hayes (Editor). Lippincott Williams & Wilkins Publishers; 2nd edition. ISBN: 0397516843 (July 2003).

Biomechanics: Mechanical Properties of Living Tissues. Yuan-Chen Fung. Springer-Verlag; 2nd edition. ISBN: 0387979476 (January 15, 1993).

EXERCISES

10.1 The function $f(x) = x + 2x^2 + x^3$ describes a velocity profile in a biofluid mechanics problem. It is of interest to simply describe what happens for $1 \leq x \leq 3$. Show that the first-order Taylor Series approximation (i.e., $f(x) \approx f(x_o) + f'(x_o)[x - x_o]$) to use in this case is $f(x) \approx 21x - 24$.

10.2 The function $f(x) = x^2$ is to be approximated at $x = 3$. What is the error between the actual value and approximate value when a first-order approximation is used and $x = 3.3$?

10.3 The relationship between biomechanical power consumption for some task and variables x and y is

$$P = 10x^2y + 5y/x.$$

Show that the optimal performance is given by $P^* = 12y$.

10.4 An adult and a child walk a mile in the same amount of time. L_1 (adult) = 1 meter and L_2 (child) = 0.5 meter. They optimize stride lengths to minimize how tired they get. By what factor will the child have to take more steps than the adult to complete the mile?

10.5 A 185 lb person whose leg length is 1 m walks at a rate of 3 km/h. Assuming that $\alpha = 0.1$ and $\beta = 0.05$, answer the following:

(a) What is the power consumption in watts when the stride length is 0.1 m?

(b) What is the power consumption in watts when the stride length is 1 m?

(c) What is the best stride length and minimum power consumption?

10.6 Two hikers are in the midst of the 100 Mile Wilderness section of the Appalachian Trail in Maine. They got separated when one hiker made a side trip to a water source and got lost (i.e., an adapted scene from *A Walk in the Woods*). Realizing they are separated, they now seek to hike out and rejoin at an agreed upon emergency rendezvous point. Each hiker has the same amount of food. Lost hiker-1 is tall and lanky while hiker-2 has a more compact build, as follows:

| Hiker-1 | M = 160 lb | L = 41 in |
| Hiker-2 | M = 180 lb | L = 36 in |

(a) One scenario is that each hikes roughly at the same speed, but to save energy, each hiker unconsciously optimizes their stride length. Based solely on the power consumption involved with walking to the rendezvous point, which hiker may run out of food first due to their different "biomechanical construction?"

(b) By what factor should hiker-2 adjust his speed to make his power consumption equal to hiker-1?

10.7 You are backpacking. Your pack weighs 20 percent of your weight. You desire that your power consumption be the same as when you walk without a backpack. You also pace yourself so that you do not burn out over a long day. When carrying the pack, by what factor does your walking speed compare to the speed that you would walk without the pack?

10.8 Three connected subsystems each possess an efficiency of 85 percent.

(a) What is the overall efficiency of the entire system?

(b) How much will the overall efficiency increase if each subsystem efficiency is increased to 90 percent?

10.9 A battery can provide enough power to a heart assist device to allow for 10 hours of mobility. How much more additional mobility (or safety factor) is obtained when the overall efficiency of converting electrical energy to blood pumping work increases from 90 to 95 percent?

10.10 The effect of gravity on the Moon is one-sixth that on the Earth. Consider the prospect that people actually live on the Moon, and while we are being speculative, assume that the Moon's gravity is one-ninth that of the Earth's. Also assume that Moonlings are twice as massive and four times taller than the average Earthling.

(a) For a given walking speed, will an Earthling astronaut or a Moonling do less, equal, or more work when taking a stroll on the Moon, and by what factor? Be sure to state your assumption(s).

(b) An average Earthling has a mass of 70 kg. How much extra mass in kg could an Earthling carry on the Moon and still have the same minimum value of power consumption as on Earth when not carrying the extra mass (i.e., Earthling + Extra Mass on Moon = Earthling on Earth; Extra Mass = x)? Be sure to state your assumption(s).

10.11 The human heart provides about 1 to 10 watts of power to circulate blood through the body. A 9.6 volt, 2.6 amp-hour rechargeable battery is used for some cordless electric drills. For 100 percent efficient conversion of electrical energy into blood pumping work, estimate how long one of these rechargeable batteries would last when 5 watts of power is consumed. Recall that 1 J/sec = 1 W = 1 A * 1 V.

Biofluid Mechanics

11.1 | Purpose of This Chapter

Fluid mechanics is a subject that deals with the properties and movement of fluids when they are under the influence of a force. For a chemical engineer, one question that typically needs to be answered when designing a process is "How big should a pump be to convey a fluid from one place to another in a reasonable amount of time and in an economical manner?" The movement of air around an airplane wing is of keen interest to a mechanical engineer because the lift forces generated keep the craft aloft. Because cells and mammals are 70 percent water, humans contain 5 to 6 liters of blood, and the continued pumping of our hearts is essential to survival, it is not surprising that fluid mechanics is also of interest to many bioengineers. Indeed, fluid mechanics is a useful basis for appreciating how we function, as well as for designing artificial organs and other devices that can help maintain a patient when things go wrong (e.g., cardio-pulmonary bypass, Figure 0.3).

This chapter will first introduce you to the basic ideas of fluid mechanics and blood's flow properties. Thereafter, these basics will be used to explain how the blood circulation system within our bodies is "designed." How fluid mechanics allows bioengineers to pursue the design of drug infusion pumps and artificial hearts will then be discussed using the basic concepts. In the case of drug infusion devices, you will see that when a designer does not fully consider the influence of pressure on materials and fluids, problems can occur.

11.2 | Mechanics of Fluid Flow

Stress and strain. Many types of forces can act upon a material's area. A **normal force** acts perpendicular to a surface. The upward-directed force a floor exerts on a standing person is an example of a normal force. A force that acts upon and along a surface is a **shear force**. A locked-up tire skidding on pavement experiences substantial shear forces. The friction between the tire and the pavement develops a force that is directed against the direction of the automobile's movement. This force also acts on the area of contact between the tire and pavement. The car will eventually stop, but the shear forces may have been large enough to peel some rubber off the tire and deposit it on the pavement. The symbol τ is often used to represent a shear force. The units of τ are force per area [e.g., 1 Newton/m^2 = 1 N/m^2 = 1 **Pascal (Pa)**].

When a shear force is applied to a material, deformation can occur, as shown in Figure 11.1. A plate is placed on the top surface of the block of material. Adhesion between the plate and the material occurs, as well as between the material and the lower surface. Pulling the plate to the right introduces a shear force that causes a deformation of magnitude, Δx, along the x-axis towards the right.

Intuition suggests that the larger the shear force is, the greater the extent of deformation. The magnitude of the deformation should also be referenced to a length scale over which the deformation is transmitted (Δy) to put into perspective how significant the overall deformation is. It makes sense to use this perspective because the shear force is also put into perspective by using a per area basis. The ratio of a deformation to a reference dimension, $\Delta x/\Delta y$, is often called a **strain**. Finally, the accumulated strain should depend on the duration of the shear force. Putting these ideas together suggests that

$$\tau\,\Delta t \sim \Delta x/\Delta y. \tag{11.1a}$$

$$\tau \sim \frac{\Delta x/\Delta t}{\Delta y} \sim \Delta v/\Delta y \sim dv/dy \sim \dot\gamma. \tag{11.1b}$$

Equation 11.1b states that the shear force is proportional to the **strain rate** ($\dot\gamma$) or the **velocity gradient**, dv/dy. This relationship makes sense from two standpoints. As long as a shear force is applied, strain should continuously occur; hence, it makes sense to have a velocity associated with reference points in an object. Secondly, if a shear force is weak, then the rate at which one point in a material moves relative to another should be low.

Fluid properties and behaviors.   We can reasonably expect that different materials will strain differently in response to the same shear force. For a fluid, viscosity (μ) provides a measure of resistance and allows for the proportionality between shear stress (τ) and strain rate in Equation 11.1b to be expressed as an equality:

$$\tau = \mu\,dv/dy = \mu\dot\gamma. \tag{11.2}$$

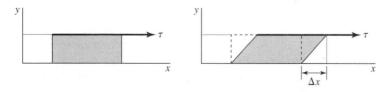

FIGURE 11.1   A shear force (τ) acting on the top area can deform the square object from its initial shape (left) to one that is strained (right).

The unit typically used for the viscosity of fluids such as water and blood is **centipoises (cP)**. One centipoise is equal to 10^{-2} g cm^{-1} sec^{-1}. At room temperature, water and air have viscosities equal to 1 and 0.018 cP, respectively. Another common unit is Pa-s, which equals 10^3 cP.

A fluid that exhibits a value of viscosity that does not depend on shear force or strain rate is called a **Newtonian fluid**. Water is a Newtonian fluid. A fluid like ketchup, however, behaves differently than a Newtonian fluid. After the bottle is inverted, ketchup initially typically resists flowing. However, after shaking, ketchup flow commences. Thus, ketchup exhibits a yield stress. Apart from exhibiting a yield stress, the apparent viscosity of some fluids changes once movement has begun. Fluids that flow more readily after motion has begun are called **shear thinning**. Others that resist movement once flow starts are known as **shear thickening**. Interestingly, high viscosity and yield stress are attributes for ketchup. Without those properties, ketchup would not cling to french fries.

Shear thinning and thickening fluids are usually macromolecule- or particle-containing solutions or mixtures. These fluids acquire their interesting properties from the different interactions that occur between their components, which flow may disrupt or foster. For example, small aggregates may form when a fluid has been at rest in a beaker for some time. When one tries to stir such a fluid, initially quite a bit of effort is required. As one persists, the shear forces imposed break up the network of aggregates with the result that stirring becomes easier. The study of fluid properties is called **rheology**. Because biological fluids contain proteins and cells, and understanding their flow properties is important for artificial organ design and other applications, an active **biorheology** research community exists.

Pressure forces.   Pressure is a force divided by the area over which it is exerted. Your body force divided by the total area of your shoes is an example of the pressure you exert on a floor. There are many units used for pressure. A common unit is the pascal (Pa). Because we all experience atmospheric pressure daily, this pressure is also a convenient yardstick to use for quantifying pressure values. One atmosphere of pressure equals about 10^5 Pa.

The heights of dense liquids are also used to express the value of pressure. Mercury, which is a dense liquid, was often used in early pressure measuring instruments. One simple instrument is a U-shaped tube that is partially filled with mercury. When pressure is exerted on one side of the U-tube, the mercury column will rise on the other side. The height the mercury rises is proportional to the pressure source measured. This traditional pressure-measuring instrument is not commonly used anymore, but the history and terminology remain in our vocabulary. Weather reports often report atmospheric pressure in mm or inches of mercury. Blood pressure measurements such as "120 over 80" also use mm of mercury as the pressure scale. One atmosphere is equivalent to 760 mm of mercury.

Because the atmosphere is always exerting pressure, we sometimes differentiate between **absolute** and **gauge pressure**. To illustrate, a diver experiences the pressure of the water above him plus that of the atmosphere bearing down upon the water. So the absolute pressure experienced is the water pressure plus atmospheric pressure. The gauge pressure is the amount of pressure that exceeds the atmospheric contribution. It is called gauge pressure because a zero reading is set at atmospheric pressure so that deviations from atmospheric pressure can be readily measured.

Finally, note that pressure and shear force have the same units because fundamentally they are both forces per unit area. It is important to not confuse them because they describe different forces on a fluid. Shear forces act parallel to physical and reference surfaces and arise due to frictional effects. Pressure forces are due to gravitational and other forces.

Mechanics of fluid flow.   Consider a fluid initially at rest in a tube. Suddenly, the pressure (P) is made higher on one end of the tube than other. The pressure difference imposed will cause the fluid to flow toward the end with lower pressure. In this case, pressure and force are exerted at both ends of the tube, but the greater force on one end results in a net force and motion of the fluid mass. The force diagram is shown in Figure 11.2.

A simple thought experiment tells us that for constant motion of fluid through a tube, two things are required: (1) a sustained pressure difference and (2) continually supplied new fluid to replace the fluid that leaves the tube. A simple illustration of how these two requirements can be met is shown in Figure 11.3. The top surface of the water in the elevated tank and the water exiting the tube both have atmospheric pressure pushing on them. However, elevating the tank raises the pressure on the water on the left side of the tube to a value that exceeds atmospheric pressure. The amount the pressure is elevated above atmospheric pressure equals the force imposed by the weight of the elevated water divided by the tube's

$$P_1 = F_1/A \longrightarrow \qquad [A] \qquad \longleftarrow P_2 = F_2/A$$

FIGURE 11.2   Unequal force (F) applied across the cross section of area A results in a pressure ($P_1 - P_2$) difference along the axis of the tube. The pressure difference will cause the fluid to move to the right if P_1 exceeds P_2.

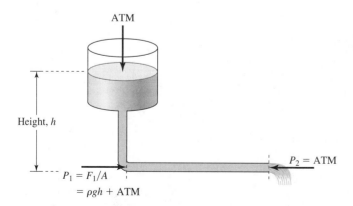

FIGURE 11.3   Example of how flow of a fluid is sustained and how a pressure difference along the axis of a tube can be generated.

cross-sectional area (A). This force is

$$F/A = mg/A = \rho Ahg/A = \rho gh, \qquad (11.3)$$

where ρ refers to the water's density [mass/volume]. Thus, water will continually be provided by the tank, and the flow through the tube will be fostered by a pressure difference that is proportional to the height the water in the tank is above the tube entrance.

This thought experiment agrees with our experience, but it raises the question "Why is a net force, or equivalently a pressure difference, needed to move a fluid?" A moving boundary can impose a shear force and move fluid elements, akin to the situation in Figure 11.1. Fluid moving past a stationary boundary is also exposed to a shear force. However, because the boundary is stationary, the force tends to defy fluid motion. To sustain fluid motion, a finite net force, provided by a pressure difference, is thus needed to overcome the shear forces induced by the presence of the boundary and the cohesive forces between fluid "particles."

The fluid closest to the boundary experiences the greatest shear force. One biomedical implication of this facet of fluid mechanics is that red blood cells can be exposed to different forces depending on where they are located in a tube or vessel and how fast the fluid is flowing. Excessive shear forces can distort or destroy deformable cells. Thus, our bodies are designed to manage these forces. Moreover, the design of artificial hearts and other devices must take these forces and exposure times into consideration. Otherwise, extensive damage to blood cells may occur.

Another consequence of the shear force declining away from a boundary is that fluid speed depends on the distance from a boundary. Figure 11.4 illustrates this outcome. The pressure drop exerts a force that pushes all the fluid "particles." The greater opposing frictional force near the boundary results in a lower net force for pushing the fluid in that vicinity. Motion is thus slower near the boundary, resulting in a spatial dependence for velocity, which is often called a **velocity gradient**. Therefore, if a red blood cell were located adjacent to a vessel or tube wall, it would have the slowest speed and be exposed to the highest shear force. The presence of a velocity gradient also explains why fish are still in a river after a flood, as opposed

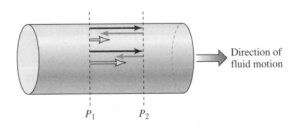

$P_1 \qquad P_2$

FIGURE 11.4 How a velocity gradient develops. When P_1 exceeds P_2 the fluid moves toward the right, driven by the force ($\rightarrow$) provided by the pressure difference. Shear stress $\longleftarrow$ induced by the boundary opposes the force provided by the pressure difference. Opposition is greater near the boundary (tube wall), resulting in less net force $\Rightarrow$ and slower fluid motion near the boundary.

to being washed away. By hunkering down on the creek bottom, they can wait out the flood in the slower-moving water.

Types of flow.   Watching water move through a slow-moving channel, as opposed to swirling about at the base of a waterfall, suggests that it may be possible to categorize different regimes of fluid flow. Indeed, if we tossed a neutrally buoyant tracer in the water, in one case it would travel on a straight line whereas in the more energetic case, the tracer would take a ragged course and perhaps periodically dwell in a swirl or eddy.

A moving fluid possesses kinetic energy; the amount per unit volume is $\rho<v>^2$, where ρ and $<v>$ are fluid density and characteristic velocity, respectively. Tending to distort and retard the motion of a fluid are shear forces of magnitude $\mu v/D$, where μ and v/D are viscosity and a characteristic velocity gradient over a representative dimension (e.g., tube diameter), respectively. The ratio of the two quantities, $\rho v D/\mu$, is called the **Reynolds number (Re)**.

If a fluid has a lot of kinetic energy (and momentum) and thus a high value of Reynolds number, it is less likely to be swayed by the presence of boundaries. Close to the boundary, shear forces will, however, create a velocity gradient. The motion of different parts of the stream can be influenced differently, with the result that the flow pattern is irregular. Such flow is called **turbulent**. Typical features of turbulent flow are swirls and eddies. When the kinetic energy is lower, shear forces dominate and the flow is smoother. This condition is referred to as **laminar flow**. For flow in tubes, the transition for laminar to turbulent flow occurs when the Reynolds number exceeds around 2,000.

Pressure difference and power requirements.   The faster a fluid moves through a tube, the greater the gradient of velocity will be, because fluid is essentially stopped at the boundary, while away from the boundary the fluid velocity must increase to a substantial value. According to Equation 11.2, the greater the velocity gradient (dv/dy), the greater is the shear force. Consequently, the pressure drop required to sustain flow is expected to increase in proportion to fluid speed.

The surface area of the boundary relative to fluid volume should also influence the total friction, and thus the pressure drop required to sustain flow, because the boundary surface is a source of shear force. For a tube, the ratio of surface area to volume is inversely proportional to radius, R. Thus, as radius decreases, we expect that the resistance to flow will increase. This expectation agrees with the experience of anyone who tries to drink a milk shake through a small-diameter straw. Using a larger-diameter straw allows one to drink faster when the same pressure difference between the straw ends is imposed through suction.

The pressure drop associated with fluid flow in a tube should thus depend, in part, on volumetric flow rate (Q; volume/time) and tube geometry. When the velocity is low and the flow is laminar (Re < 2,000), the **Hagen-Poiseuille** law provides the dependence of pressure drop (ΔP) on volumetric flow rate (Q), fluid viscosity (μ), and tube geometry (i.e., radius R and length L). The velocity and shear stresses generated can also be found by the analysis that leads to the Hagen-Poiseuille law. The relationships are

$$\Delta P = 8\mu L Q/\pi R^4. \tag{11.4a}$$

$$v = \frac{\Delta P R^2}{4\mu L}[1 - (r/R)^2]. \qquad (11.4b)$$

$$\tau = \Delta P r/ 2L. \qquad (11.4c)$$

It is helpful to see whether the above equations' predictions agree with intuition. Equation 11.4a states that the higher the flow rate or the smaller the radius of the flow path, the greater the pressure drop generated or needed. Equations 11.4c and 11.4c together indicate that near the wall ($r = R$), the stress is greatest and the velocity is the lowest. These predictions agree with the experiences of blowing air through straws of different diameters and the fish resting on the bottom of a fast, flowing stream.

Because fluid flow requires a force, and a continual physical displacement of mass occurs, work is continually performed. The rate work is performed (dW/dt) equals the product of force (F) and characteristic velocity ($<v>$); hence, the power required to move a fluid is

$$\text{Power} = dW/dt = F<v> = A\,\Delta P<v>. \qquad (11.5)$$

In a tube of known cross-sectional area (A) and length (L), the rate at which fluid volume is ejected is

$$Q = (AL)<v>\,/L = A<v>, \qquad (11.6a)$$

which allows us to determine the power required as

$$\text{Power} = Q\,\Delta P. \qquad (11.6b)$$

Equation 11.6 is a very important relationship for engineering work because it answers the bottom line question, *"How much power is needed to get a fluid-moving job done?"* The difficult part sometimes is to determine the pressure drop associated with a particular flow rate and geometry, which is an input in the power calculation. When the flow is laminar and in cylindrical geometry, and thus obeys Equation 11.4a, the pressure drop calculation is straightforward. When other types of flow occur, other analyses or even computer simulation are required. Equation 11.4c finds use in estimating the stresses that may be imposed on objects such as cells.

11.3 | Blood versus Water

There is an old saying, "Blood is thicker than water." While meant to describe relationships between relatives, there is some rheological truth to the saying. About 45 percent of the volume of blood is made up of hemoglobin-containing red blood cells. These cells impart the characteristic red color to blood and the ability to carry oxygen to tissues. Red blood cells have a disk-like shape, where the center is thinner than the outer periphery. A range of shapes that deviate from the disk-like geometry can be found in blood samples, which may relate to cell age and history. Typical and

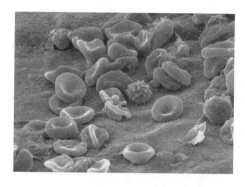

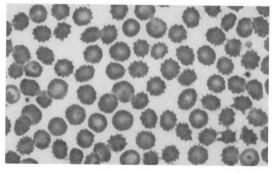

FIGURE 11.5 (Top, left) Micrograph of typical red and white blood cells in a vessel. (Top, right) The shape of a sickle cell differs from that of a typical, disk-shaped red blood cell. (Bottom) The altered morphology of crenated red blood cells.

atypical red blood cells are shown in Figure 11.5. This volume fraction of red blood cells is called the **hematocrit** and it is commonly used as a diagnostic index. A low hematocrit, for example, is indicative of anemia. The remaining 55 volume percent of blood is a solution of water, salts, and proteins. The fluid devoid of red and other cells is commonly called **plasma**.

Like its major constituent, water, plasma behaves like a Newtonian fluid. The viscosity of plasma is around 1.2 cP. When red blood cells are added, however, the rheology can become more interesting. Blood tends to exhibit a yield, beyond which it behaves as a shear thinning fluid. This means as shear stress is increased from zero, blood will not immediately flow. Rather a **yield stress** (τ_Y) exists and must be exceeded by the imposed stress for flow to occur. As shear stress is increased further, the apparent viscosity decreases and then levels out to 3 to 4 cP, which is still higher than plasma or pure water. The non-Newtonian behavior of blood is due to the presence of red cells, so it stands to reason that varying the hematocrit (H) will alter the apparent viscosity of blood. Numerous relationships have been found to describe the dependence of yield stress on blood composition, and also the relationship between shear stress (τ in dynes/cm^2) and strain rate ($\dot{\gamma}$ in sec^{-1}) The **Casson equation** is one example:

$$\tau^{1/2}{}_Y = (H - 0.10)(C_F + 0.5). \tag{11.7a}$$

$$\tau^{1/2} = \tau^{1/2}{}_Y + s\dot{\gamma}^{1/2}. \tag{11.7b}$$

$$s = [\mu_o/(1 - H)^{\alpha a-1}]^{1/2}. \tag{11.7c}$$

In this model, C_F, μ_o, and $a\alpha$ refer to fibrinogen concentration (g/100 ml), plasma viscosity, and a protein composition-dependent parameter, respectively. Additionally, s corresponds to the square root of the apparent (Newtonian) viscosity that is exhibited at high values of $\dot{\gamma}$. Equations 11.7 (a) and (c) both capture the behavior of how yield stress and high-strain viscosity (s^2) increase with hematocrit. Equation 11.7b fits data that show that τ does not increase linearly with $\dot{\gamma}$, as it would with a Newtonian fluid.

When the viscosity is constant and the flow is laminar, Equations 11.4 (a-c) work well. Because the apparent viscosity of blood is variable, these equations can provide approximate answers or some qualitative insights at very low flow rates. When the shear stress is higher and the apparent viscosity of blood has "leveled out," Equations 11.4 (a-c) can provide some reasonable quantitative results. Some examples follow to show how to use the equations, deal with different units, and glean additional insights on biofluid mechanics.

11.4 Example: How Much Force Is Needed to Inject a Drug?

To illustrate the use of the basic equations for laminar flow, consider the design of a hypodermic syringe. A small flow channel (i.e., narrow diameter needle) is used to inject a drug because it is undesirable to make a large puncture in the patient, which could result in wounding and infection. It is also easier to penetrate the skin with a modest force because the small cross-sectional area of the needle transmits the modest force applied into a significant puncture pressure. The small flow channel, however, increases the pressure drop required for the drug to flow out of the syringe in a timely manner. If the force required is high, the person administering the medication may not be able to achieve a fast flow, thereby prolonging the unpleasant experience of getting a shot in the arm or elsewhere.

Let's assume that 4 cc of a drug (V_D) needs to be injected in at least 2 seconds (t_D). This corresponds to a flow rate of $2*10^{-6}$ m³/s. Because viscous forces are high, let us also presume that the flow is laminar. To simplify the calculation, we will also assume that the "back-pressure" imposed at the needle's end within the body is similar to the background pressure that the atmosphere exerts on the plunger. Thus, the pressure drop associated with drug flow equals the pressure that must be generated by the person performing the injection. As shown in Figure 11.6, the drug, whose viscosity is 3 cP ($3\ 10^{-3}$ kg m^{-1} s^{-1}) and density is 1 g cm^{-3} (10^3 kg/m³), will flow through a $5*10^{-2}$ m long (L), 20 gauge needle with a bore radius (R_b) of $3.3*10^{-4}$ m. The radius of the plunger (R_p), where the force for fluid motion is obtained through pressure exerted by the thumb, is $5*10^{-3}$ m.

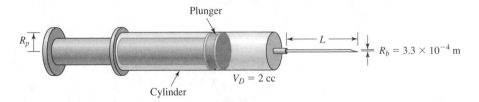

FIGURE 11.6 Geometric parameters for a syringe; the subject of a calculation example in the text.

First, we determine the value of the Reynolds number in order to double-check our assumption that the flow within the bore is laminar and find that, indeed, Re is less than 2,000:

$$Re = \rho v D/\mu = \rho Q\, 2R_b/\pi R_b^2 \mu = 2\rho Q/\pi R_b \mu = 1{,}286 < 2{,}000.$$

Next, we need to determine the pressure drop required for the 2 cc of drug to flow in 2 seconds. The pressure drop could be calculated with Equation 11.4a, and then the force determined by multiplying the value of ΔP by πR_p^2. However, it is often useful to keep an analysis general, and plug in the numbers in the final step. Often when an analysis is kept general, some problem details cancel out and the key variables and relationships that pertain to a particular question can be readily isolated.

The radius of the needle's bore is much smaller than the radius of the cylinder. Therefore, a reasonable assumption is that flow through the needle's bore accounts for the majority of the pressure drop. The general result obtained from using Equation 11.4a is

$$F = \Delta P \pi R_p^2 = 8\mu L\frac{V_D R_p^2}{t_D R_b^4}.$$

The force required increases rapidly as R_b decreases. Thus, using a smaller dimension needle to inflict less injury and use less penetration force trades-off against the need for increased force to achieve drug flow. The force needed to prompt drug flow also increases linearly with injection volume and inversely with injection time.

Using units of m, kg, and seconds yields a value of force equal to 5.06 N. This is equivalent to 516 grams of mass in Earth's gravitational field, which corresponds to about 1.5 cans of soda. Because the friction between the inner cylinder wall and the plunger was not considered, 5.06 N is a minimum estimate of the force needed.

11.5   Example: How Does the Heart Compare to a Lawn Mower Engine in Horsepower?*

The five liters of blood in a person circulates roughly every minute, which corresponds to a flow rate of $2.9*10^{-3}$ ft^3/second. The pressure drop in the circulatory system is on the order of 100 mm mercury. The pressure equivalent of this column of mercury is 278.5 lb$_f$/ft^2. The power is given by Equation 11.4c. Noting that one horsepower is 550 ft lb$_f$ s^{-1}, we find that

Power = $2.9*10^{-3}$ ft^3/s * 278.5 lb$_f$/ft^2 * 1 Horsepower/550 ft-lb$_f$ s^{-1}

= 0.00149 Horsepower.

Compared to the average five horsepower lawn mower, there are 3,356 human hearts per lawn mower. The human heart uses much less power than a lawn mower. One reason is that despite the long and intricate system of small diameter blood vessels, the pressure drop is fairly low. One reason for the lower than expected pressure drop is provided in the section following the next example.

*Adapted from D.O. Cooney's text, "Introduction to Biomedical Engineering," pages 84–85.

11.6 Example: What Is the Stress on a Red Blood Cell?

A patient on an operating table is to be sustained with a blood bypass, heart-lung machine. Here, the patient's blood is pumped through a device that oxygenates the blood and then returns the blood to the patient to keep tissues alive during a surgical procedure. Such a device was briefly described in Chapter 0 and can be reviewed in Figure 0.3. Assume that one option is to use a 1-meter long tube with a 2 cm radius to connect the patient to the machine. If the patient's blood is passed through the tube to the machine at a rate of 5 liters/minute, what is the shear force that the red blood cells experience in the tube, *and* is the force significant for this option?

To answer these questions, a general relationship is first obtained. Combining Equations 11.4a and 11.4c, the shear stress when $r = R$ is

$$\tau = \frac{8\mu L Q \, r}{\pi R^4 \, 2L} = \frac{4\mu \, Q}{\pi \, R^3}.$$

A number of variables cancel (e.g., L) and note that $r = R$ was chosen to find the maximal value of shear stress, the stress at the tube wall. We sought the maximum force in order to find the worst-case scenario, because we prefer to make decisions that are based on the highest reasonable degree of caution. The distilled result tells us the faster the flow rate (Q), the greater the stress. Additionally, the shear stress will increase dramatically as the tube radius is decreased; every halving of radius will increase the wall shear stress by eight-fold.

Using the values of flow rate ($Q = 8.33*10^{-5}$ m^3 s^{-1}), viscosity ($\mu = 3*10^{-3}$ kg m^{-1} s^{-1}), and radius ($R = 2*10^{-2}$ m), the stress is 4 Pa. The asymptotic value of blood viscosity was used to simplify the analysis. An alternate way of viewing this assumption is that we are assuming that the strain rate ($\dot{\gamma}$) is high enough so that blood flow is essentially Newtonian with an apparent viscosity of 3 cP (see Equation 11.7b). The exposure time in the high shear stress region exceeds 15 seconds, which is the average time the blood stays in the tube (from Equation 11.4b, $t = L/<v> = LA/Q$.)

Is this 4 Pa shear force significant? To answer this question, we need to know the mechanical properties and response of red blood cells. When exposed to a shear flow, red cells become elongated and otherwise distorted due to the net shear forces acting on them. The elastic properties of cell membrane-skeletons has been measured (e.g., Lenormand, G., Henon, S., Richert, A., Simeon, J., and Gallet, F., 2001) and attempts have been made to directly observe the distortions of cell shape when exposed to defined, shear flows such as those that can be established in capillaries and devices known as rheometers (Baskurt, O., Gelmont, D., and Meiselman, H., 1998). Additionally, how long it takes a red cell to recover its shape, if it can, after being exposed to a particular value of shear stress for a given time period, has been investigated (e.g., Markle, D., Evans, E., and Hochmuth, R., 1983).

Shear stresses in the range 0.5–50 Pa have been found to distort red cell shape when cells were suspended in a viscous medium to limit their tumbling (Baskurt et al., 1998). The magnitude of the shape distortion tended to level off after 10 Pa. Crenated cells, which means the cells were treated to change their disk-like shape into a sphere (recall Figure 11.5, bottom), exhibited elongation when bound to a glass surface and exposed to a shear stress equal to 0.6 Pa. Flows in capillaries that imposed wall shear stresses of 3 Pa were sufficient to distort human and avian (i.e., bird) red cells (Gaehtgens et al., 1981).

Prior work suggests that the 4 Pa wall shear stress is significant in that it could distort the shape of a red cell. Moreover, how long such a force is exerted is important because excessive duration could lead to permanent damage. Overall, exposure to shear forces and shape distortion are natural aspects of red cell function that allow them to wiggle through the narrow blood vessels. However, bioengineers are cognizant of the fact that too much stress for too long a time can exceed the ability of red cells to remain intact and functional. Thus, substantial research has been done on the mechanical properties of red blood cells. Stress minimization is a key objective of bioengineering design conceptualization and calculations. Examples of life support systems and artificial organs and their design issues are provided at the end of this chapter.

11.7 Operation and Design of the Circulatory System

Pulsatile flow & vessel compliance.   The chamber design of the human heart was described in Chapter 9. The squeezing action of the ventricles and the opening and closing of valves accounts for the noises the heart makes and why a physician can infer how a patient's heart is functioning by listening with a stethoscope. The periodic contractions also cause the blood flow to be uneven. Flow acceleration and deceleration occur in response to the atrium filling and ventricle contraction rhythms. Such periodic flow is termed **pulsatile flow**.

Some of the peaks and valleys in blood flow rate are flattened due to the elastic nature of the aorta, capillaries, and other blood vessels. For example, a left ventricle contraction imposes a pressure pulse on the circulatory system. In response to elevated pressure, the aorta and other downstream vessels stretch, akin to inflating a balloon. The distortion of shape is known as **compliance**. The elastic aorta and other vessels absorb some of the force imposed much like the way a coil spring cushions the ride in a car when bumps are encountered in the road. The stretching allows for flow to occur at a lower pressure because the effective vessel radii increase.

The stretched aorta (and other blood vessels) can also store some energy in the manner a stretched spring can store energy. When the contraction is over, the energy stored in the stretched vessels is released and the vessels' radii decrease. The blood-squeezing action of contracting blood vessels helps to keep the blood moving during the interval between ventricle contractions. The material properties of blood vessels significantly influence what pressure is experienced at different locations in the circulation. Interestingly, far downstream, the hills and valleys in pressure and flow tend to be flattened out by compliance, with the result that blood flow is fairly constant.

Arterial and venous circuits.   The nature of the vessels and blood flow in the body was elucidated by Galen, Harvey, and other pioneers as documented in the

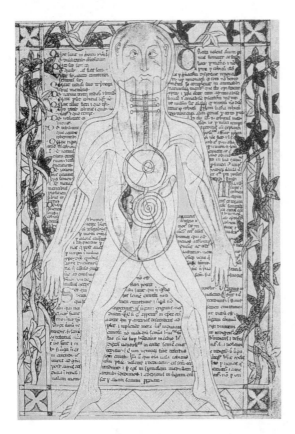
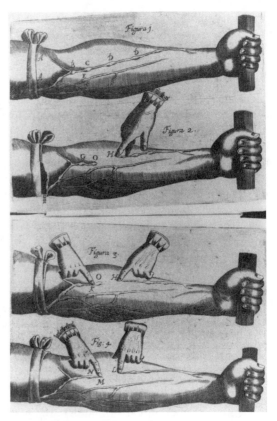

FIGURE 11.7 (Left) An illustration of the venous system created in the late 13th century. Galen found that the venous blood flow was separate from the arterial blood flow, and blood flow was pulsatile. (Right) The simple but illuminating experiment performed by William Harvey, which he reported in *On the Circulation of the Blood* (1628). Harvey demonstrated that the blood in the veins flows in one direction and to the heart. An arm was bound to raise the veins and thus make them more visible. Blood in one vein was then squeezed away from the heart. The vein did not refill because backflow from the heart is blocked by valves.

artwork from the 1300s and 1600s (Figure 11.7). Recall from Chapter 9 that **arteries** are the conduits for blood leaving the heart and entering organs. **Veins** are the conduits that return blood from organs back to the heart. The arteries branch into smaller vessels until the smallest vessels, **capillaries**, are reached. There are 25,000 miles or more of capillaries in an adult and if all the blood vessels in an adult were connected end-to-end the total length would exceed 60,000 miles (sec: American Heart Association, "Heart, How it Works," http://www.americanheart.org/presenter. jhtml?identifier = 4642). Because the pressure drops as one goes downstream in a flow circuit (recall Figures 11.2 through 11.4), the pressure is higher in arteries than veins. Consequently, arterial walls are thicker than venous walls, to limit the distortion in shape that fluctuating pressure can induce.

Capillary branching. There are advantages associated with using small-diameter capillaries. As the radius decreases, for example, oxygen has a smaller distance to travel from the red blood cells dispersed within the capillaries to the adjoining tissues that require oxygen. Thus, small-diameter capillaries enable the oxygenation of tissues. The advantages provided by small-diameter capillaries, however, raise the fluid mechanics-inspired question: "*Based on Equation* 11.4, *why is the blood pressure not extremely high when blood is pumped through a network of small diameter vessels that enable tissue oxygenation?*" The explanation lies in how the circulatory system is designed, and in the properties of blood. Considering the design of the system, the branching of vessels is extensive as one goes further downstream from the heart. This means that the total flow rate gets divided into **parallel flow circuits**.

The advantage of branching can be illustrated by viewing fluid flow in terms of Ohm's law and an electrical circuit. Ohm's law is a general statement: it says that the product of flow rate (e.g., electrical current, i) and resistance (e.g., circuit resistance to current flow, r) equals the potential (e.g., volts, v) needed to drive the flow. The Hagen-Poiseuille law can be cast into the same form as Ohm's law where potential, current flow, and resistance physically correspond to ΔP, Q, and r_f, respectively:

$$v = ir. \tag{11.8a}$$

$$\Delta P = 8\mu LQ/\pi R^4 = Q[8\mu L/\pi R^4] = Qr_f. \tag{11.8b}$$

The pressure drop for two different circuits, unbranched and branched, can now be more easily compared. An example is shown in Figure 11.8. A vessel with total flow Q is contrasted with two vessels in parallel that have the same length and radius, but each carries half the flow. Note that the pressure drops along the two parallel vessels are the same.

Letting the subscripts "U" and "B" denote the unbranched and branched cases, respectively, the ratio of the pressure drop in the branched versus unbranched circuit $(\Delta P_B/\Delta P_U)$ is

$$\Delta P_B/\Delta P_U = 0.5\, Q\, r_{fB}/Qr_{fU} = 0.5\, r_{fB}/r_{fU} = 0.5\, (L_B/L_U)(R_U/R_B)^4 = 0.5. \tag{11.9}$$

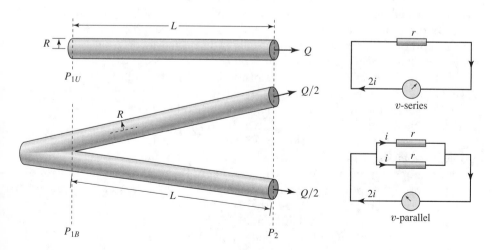

FIGURE 11.8   A unbranched versus branched flow circuit and an electrical analogy. Using parallel versus single flow paths can reduce the total pressure drop or voltage required.

Thus, replacing a single vessel with two in parallel with radii and length equal to the single vessel reduces the pressure drop needed by one-half. This outcome corresponds to an electrical circuit. When a current of value $2i$ passes through a resistance r, the voltage needed to push current is $2ir$. If instead the same total current is split and carried by two equal resistances in parallel, the total resistance of the equivalent circuit is $0.5r$. The voltage required to push $2i$ through an equivalent resistance of $0.5r$ is now reduced by half and equals ir.

An interesting design is apparent where the trade-off between ample oxygen delivery and high pressure drop is managed through the architecture of the circulatory system. Branching into smaller vessels reduces the pressure drop *while* also enhancing the delivery of oxygen and nutrients from the vessels to the surrounding tissues. However, Equation 11.9 indicates that the advantages begin to disappear when the branched vessels become long or especially very small in radius because pressure drop depends on $1/R^4$. Also, if the vessel radii are too small, the red blood cells will have to deform considerably in order to pass, if transit is even possible. You can explore how changing the relative proportions of vessel length and radius in an unbranched versus branched circuit affects the ratio of pressure drop and vessel surface area, or by using the flow calculator on the website.

11.8 Biomedical Engineering Applications, Accomplishments, and Challenges

Infusion devices. Sometimes even simple devices can fail with adverse consequences when physiology, fluid mechanics, and material properties are not fully considered together. An interesting example is the use of infusion pumps for administering medication to premature infants. An infusion pump is a positive displacement pump. In one configuration, it consists of a medication-containing cylinder. A piston is located in the cylinder's bore. As the piston is pushed into the cylinder by a stepper motor, the medication in the cylinder is ejected from the cylinder, much like the operation of a large hypodermic needle. The ejected medication is usually administered to a patient via an intravenous line so that it is directly injected into the blood stream. By altering the rate the piston is pushed through the bore, small or large doses can continually be administered in accordance with the patient's response. Some examples of infusion pumps can be explored using the links provided on the website.

A number of years ago, this seemingly simple device generated substantial problems when used for small patients such as neonates under critical care. Erratic patient responses frequently occurred, resulting in the need to more closely monitor a patient and respond rather dramatically when blood pressure or some other clinical index faltered. It turned out that the patients' erratic symptoms were not necessarily intrinsic to the medical condition. Rather, when slow pumping speeds were used, the compliance of the cylinder and tubing materials conspired to provide highly uneven flow of medication. Part of the time, little or no medication flowed. Other times, **slugs** of medication were injected into the patient.

The problem was that as the piston was slowly stepped down the bore, the incrementally increased pressure stretched the materials. Pressure built up and little medication was ejected. Pushing the piston a bit more stretched the materials a bit further. Ultimately, further stretching became difficult and additional piston motion triggered the release of energy stored in the stretched materials. The sudden contraction of volume triggered by built-up pressure resulted in a large slug of medication flowing into the patient. For small patients, even modest variations in medication flow rate can manifest as apparent underdoses or overdoses of medication and, in turn, erratic blood pressure and other clinical indices.

This problem normally did not arise when the pumps were used for larger adults. Uneven flows of medication have lesser impact on larger adults. The higher flow rates used for adults also did not encourage the stepwise stretching and sudden-release phenomenon, because the piston motion was faster and steadier. Consequently, special pumps had to be designed for infants and small children, or materials reengineered to allow for the treatment of both neonates and adults with the same equipment.

Heart replacement and assist devices.   An ongoing considerable challenge is to develop a total replacement heart for a human. Because the heart is simply a positive displacement pump, this problem may at first seem not to be formidable. However, when one considers that the superb reliability of the human heart has to be matched with a mechanical surrogate, then the engineering challenges begin to clarify. Now consider that the materials used to build an artificial heart have to be nontoxic and not engage the body's immune or wound-healing defense systems. Then add the challenge that every pump requires a power supply. The power supply must not be so cumbersome that the recipient is essentially confined to a bed or chair. Rather, mobility and quality of life are desirable rather than simply patient survival. Next, the natural heart responds to demand; when we run, blood circulation increases. Therefore, an artificial heart must also possess a control system that "knows" how fast to circulate blood to meet varying physiological needs. Finally, roughly half of the blood volume is oxygen-carrying, red blood cells. Thus, the artificial heart must circulate the blood without breaking or damaging the red blood cells through excessive shear forces and long exposure to moving parts.

Tackling these engineering problems is important because more than one million people die per year in the United States with cardiovascular disease listed as a primary or contributing reason on their death certificates. To put the number of deaths into perspective, cardiovascular disease-related deaths account for 40 percent of the total deaths per year in the United States (for more details see *Heart Disease and Stroke Statistics—2003 Update*, American Heart Association, http://www.americanheart. org/presenter.jhtml?identifier=1928 and http://www.americanheart.org/downloadable/heart/10590179711482003HDSStatsBookREV7-03.pdf). While cardiovascular disease encompasses many pathologies, a significant number of these people would have benefited from heart replacement or assist devices (for more details see *Expert Panel Review of the NHLBI Total Artificial Heart (TAH) Program*, June 1998-November 1999, http://www.nhlbi.nih.gov/resources/docs/tah-rpt.htm). However, only about 2,000 patients per year benefit from the transplantation of natural hearts from donors. Moreover, the number of donated hearts is not increasing relative to population growth; hence, there is considerable impetus for perfecting mechanical replacement and assist devices. Worldwide, the number of people with diseased hearts provides, of course, an even more compelling case for perfecting heart assist and replacement devices.

The prevalence of degenerative heart disease and trauma-induced heart injury has prompted the National Heart, Lung, and Blood Institute of the U.S. National Institutes of Health to spent millions of (U.S.) dollars on research and development since 1982, which is when **Barney Clark** became the second recipient of a total permanently-intended, artificial heart. Today's typical college student was born in the 1980s and over his or her lifetime, which is marked at the beginning with the Barney Frank story, many interesting concepts have emerged, but patients with artificial hearts tend to survive for days to months rather than years. Some in the biomedical community feel that artificial hearts should be viewed as a bridge to transplantation, while others strive to perfect a long term, total replacement.

One promising example of a total replacement heart is the **AbioCor artificial heart,** which is illustrated in Figure 11.9. The AbioCor is constructed from titanium and the custom-made, polyether-based polyurethane plastic Angioflex$^{(TM)}$. An internal motor drives the "periodic squeezing" of blood through valves, with the result that a heartbeat is simulated. Engineers are thoroughly engaged in the design and development of such systems. The veteran heart transplant surgeon O.H. Frazier aptly summarized the accomplishments of engineers in an interview about artificial hearts and the AbioCor system with Sarah Holt for NOVA:

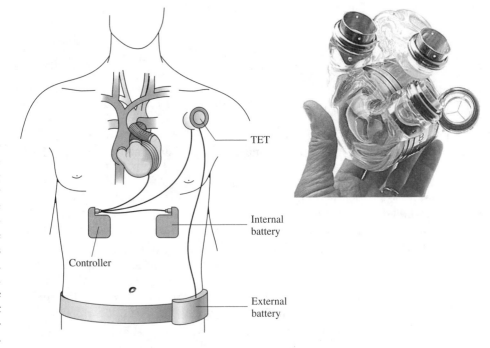

FIGURE 11.9 (Left) A sketch of an implanted AbioCor artificial heart and the power system used. Rechargeable internal and external power supplies are utilized. Power can be transmitted across the skin via electric field induction thereby obviating the need for wiring that crosses the skin, which can function as an avenue for infection. (Right) Close up of the AbioCor total replacement heart. Images used with permission from ABIOMED Inc.

TET

Internal battery

Controller

External battery

Frazier: . . . . . Another advantage is that the AbioCor pumps alternate right to left. So it doesn't pump both to the lungs and the body at the same time, like the normal heart, but it pumps side to side. This is an important advantage, because the bulk of the volume compensation that a pulsatile pump needs can be adjusted by using alternating ventricles. So when the left side is pumping, the right side is a volume chamber, and vice versa.

NOVA: Why is that important?

Frazier: Well, the right heart and the left heart don't pump the same amount of blood; the left heart pumps more. So there has to be some way of compensating for that variance between the two sides of the heart and the amount of blood pumped. This has been done in a very ingenious way by the engineers who developed the AbioCor heart—a really outstanding engineering accomplishment. They designed a small compensation chamber, which acts in a way to allow the right side of the heart to sense whether it needs to pump more blood or less blood. It automatically adjusts internally according to changes within this little chamber.

(From NOVA interview with well-known surgeon, Dr. O.H. Frazier. http://www. pbs.org/wgbh/nova/eheart/frazier2.html.)

While the quest for a total replacement heart continues to make progress, remarkable success has been achieved in the development of devices that *assist* the heart when it is damaged or weakened. These devices typically are implanted into the body to supplant the function of the left ventricle, which works harder than the right ventricle. These devices thus have become known as **LVADs**, which is shorthand for Left Ventricle Assist Devices. A well-known example is the **HeartMate**, conceived and built by Thoratec Corporation. This FDA-approved bridge to a transplant pump has an external power supply. There are air- or electric-motor-driven pumping versions. It is surgically inserted between the left ventricle and the aorta. Figure 11.10 shows what a HeartMate looks like; it is about 4 inches in diameter and 2 inches thick.

An intriguing alternative LVAD design is the **axial flow pump**. A screw-like device is mounted on a shaft, as shown in Figure 11.10. A quiet magnetic induction motor is used to rotate the screw. Blood is conveyed along the screw threads and forced out. MicroMed Technology, Incorporated is a prominent innovator and developer of axial flow LVADs. Their product is called the MicroMed DeBakey VAD®. The LVAD is named for Dr. Michael DeBakey, a well-known heart transplant surgeon. The company and device emerged from a collaboration between Drs. DeBakey and Noon and NASA engineers.

When an LVAD is implanted, the heart has to work less, because a supplemental source of energy and blood pumping is provided. Thus, when an LVAD is implanted and used over a short time period, a patient's condition can be stabilized. When implanted over a longer time, the patient's quality of life can be somewhat reclaimed. A patient can travel and engage in other mobile activities because the power supply requirements and physical size are modest. Interestingly, there have been cases where allowing the heart to rest by using an LVAD to perform left ventricle work has reversed the weakened state of the heart. Learning that one is now no

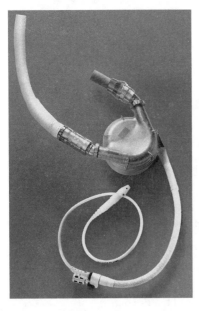
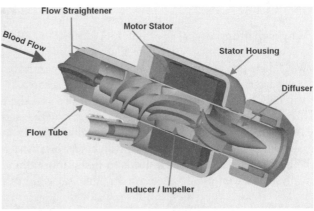

FIGURE 11.10 Implantable heart assist devices. (Top, left) The current left ventricle assist device dubbed the HeartMate XVE and built by the Thoratec Corporation, Pleasanton, CA (image courtesy of Thoratec Corp, Woburn, MA). (Top, right) Schematic of the DeBakey axial flow pump developed and vended by MicroMed Technology, Inc. (image courtesy of MicroMed Technology, Inc., Houston, TX).

longer a heart transplant candidate is not only good news for the patient and physician, but the cost of medical care is also considerably lowered. In other cases, the overall health of LVAD-recipients improved such that they were better able to cope with the stress of heart transplant surgery.

When some types of LVADs are implanted, the blood flow becomes continuous rather than circulating in the normal pulsatile fashion. Thus, some LVAD-implanted patients have no pulse. Whether a human could live without a pulse was an interesting medical and philosophical question. Apparently, it is possible to live without a pulse, as many LVAD recipients attest. One potential reason for this outcome was noted earlier. Recall that deep with the blood circulation, blood flow is less pulsatile due to the compliance of blood vessels. Thus, many tissues experience fairly even blood flow; hence, a continuous output LVAD may not necessarily disrupt normal physiology.

11.8.1 Computation and device design in complex geometries

As noted earlier in this chapter, appreciating the magnitude of the spatially-varying velocities and stresses present in a flow system is important because blood cells may be adversely affected. Insights on, for example, where stress is high are obtainable from Equation 11.4 when flow occurs through a cylindrical vessel, such as that depicted in Figure 11.4. However, when a more geometrically complex situation arises, such as the rotor of an axial flow pump, the analysis and device design problems are considerably more difficult to solve. Indeed, even though a tube is a three-dimensional object, Equation 11.4 can be obtained without great difficulty because the fluid flows in only one dimension and radial symmetry can be exploited in the derivation. Flow through an axial pump is not as simple; a three dimensional path is instead taken by the blood. The higher dimensionality and other new complexities do not bode well for performing a quick and accurate analysis of how stress and velocity vary with position, and what magnitudes can be expected. The difficulty in obtaining such information, in turn, makes it problematic to select the rotor dimensions and rotational speed that maximize device performance *while* minimizing cell damage. This section provides a glimpse of one tool biomedical engineers use to analyze blood flow in complex geometries, **Finite Element Analysis** (FEA). To illustrate the ideas behind FEA, we shall first describe the information one seeks for device design, and then describe how the information is acquired through FEA.

One strategy for optimizing the design of a blood-moving device is to first break up the moving blood volume within a device into small volumes. For example, the volume element could be a cube with the side-dimension Δx. The product of stress and characteristic dwell time (i.e., $\Delta x/<v>$) can now be computed for each volume element. This multiplication is relevant because a large value of the product is undesirable. For example, a large value means that an albeit short exposure to a high stress occurs at the cube's location. Alternately, a large value of the product could indicate that the dwell time is large at the cube's location. This alternative is also not desirable if the cube's location is proximal to a surface associated with the device. As the next chapter will discuss, the interaction between blood and a surface may lead to blood coagulation, which is highly undesirable. To complete the evaluation of the design, it is reasonable to sum the products from each volume element and then compute an average value. Alternately, "hot spots" can be identified where some locations within the device exhibit large values of the product. A high value of the average or the occurrence of many "hot spots" may prompt the designer to alter the device's shape in order to minimize the average value of the product or the frequency of "hot spots." Metrics other than the "product" just described could be developed that express an analyst's design goals and different assumptions on what most impacts blood cells.

The above optimization assumed that one had the values of stress and characteristic fluid velocity for each volume element. In complex geometries, FEA can provide these desired values. When performing FEA, the volume is first "meshed" in order to create smaller volumes. The "meshing" step is often not as straightforward as breaking a volume up into a large number of small cubes, but the intent is similar. Within a small element, it becomes possible to simplify the "exact" (differential) equations that describe fluid flow. To gain a sense of how "meshing" can be the prelude to equation simplification, recall from Chapter 10 the use of the Taylor series approximation. There, an "exact" yet cumbersome trigonometric expression was replaced with an approximation that provides satisfactory accuracy over a small span of variable value. Of course, all the elements must be mathematically "connected" to insure that the continuity of velocity and other "reality" conditions are met.

Overall, a problem that is difficult to solve by hand is broken down into elements and solved via a computer. The analyst can contrast the attributes of different meshed shapes via separate FEA calculations. Some analysts even directly link the "geometric design" to the FEA calculations. In this case, the FEA results continually and automatically drive the adjustment of the shape until an objective such

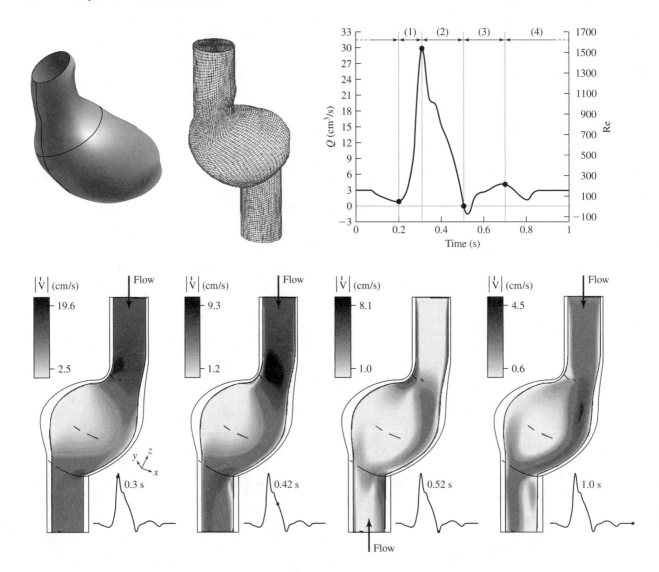

FIGURE 11.11   Example of a finite element analysis of a biomedical problem using patient-based data. One aim is to determine the spatial distribution of blood velocity in an abdominal aortic aneurysm during the cardiac cycle. (Top, left) A solid and meshed model of a particular patient's dilated blood vessel. (Top, right) The calculated dependence of the flow rate and Reynolds number in the blood vessel as a function of time during a cardiac cycle. The four flow phases shown are (1) systolic acceleration, (2) systolic deceleration, (3) early diastole, and (4) late diastole. Peak flow occurs at 0.304 s and the diastolic phase begins at 0.52 s. (Bottom) The regions of differing blood velocity at different points in the cardiac cycle. At 0.30 s, a high velocity jet of fluid is entering the vessel. At 0.42 s, the flow decelerates as it begins to separate from the proximal neck of the aneurysm and downstream at the outlet. The maximum velocity at 0.42 s equals about half the peak flow velocity, as flow recirculation is minimal and absent at the center of the aneurysm. The flow reverses its direction at 0.52 s. The pressure gradient exerted in the opposite direction (the flow now travels upward) results in a low-velocity clockwise-rotating vortex at the center of the aneurysm. At 1.0 s, the typical flow pattern that characterizes late diastole occurs: low-velocity flow recirculation at the center of the aneurysm sac, translation of the vortex center downstream at the end of the cycle, and a faster forward moving flow that shears the lateral-posterior wall as the cardiac cycle comes to an end. Used with permission from C.H. Amon, Carnegie Mellon University.

as minimized "hot spot" occurrence is reached. When one further considers that blood cells can distort in response to local stresses, mechanical models of cells can be also imbedded in such calculations to assess how cells may actually respond to a given flow environment.

FEA finds many other uses such as "mapping" the flow of blood in normal and altered vessels in order to understand better how the body is "engineered" and the impacts of injuries or disease. A meshed volume and what FEA can yield are shown in Figure 11.11. In this example, the flow of blood through an aortic aneurysm during the cardiac cycle was analyzed in order to evaluate the pressure imposed on the vessel wall and the circulation pattern of blood through the vessel.

ADDITIONAL READING

Biomechanics. Y.C. Fung. Springer Verlag; 2nd edition. ISBN: 0387943846 (November 21, 1996).

Cardiac Assist Devices. Daniel J. Goldstein (Editor) and Mehmet Oz (Editor). Futura Publishing Co; 1st edition. ISBN: 0879934492 (January 15, 2000).

Electric Heart. NOVA episode available on videotape. Sarah Holt, Executive Producer. WGBH Educational Foundation.

Life in Moving Fluids. Steven Vogel. Princeton Univ Press; 2nd Revision edition. ISBN: 0691026165 (April 1, 1996).

Transport Phenomena. R. Byron Bird, Warren E. Stewart, and Edwin N. Lightfoot. John Wiley & Sons; 2nd edition. ISBN: 0471410772 (July 25, 2001).

EXERCISES

11.1 What pressure is experienced by a diver 10 meters beneath the surface of a lake? Express your answer in Pascal units ($1 \text{ N/m}^2 = 1 \text{ Pa}$). Also express your answer in atmosphere units ($1 \text{ atm} = 10^5 \text{ Pa}$). How many meters of water are equivalent to the pressure that the atmosphere continually exerts on us?

11.2 For laminar flow in a tube, draw how velocity and shear stress vary with radius. Where are the shear stress and time spent by a fluid element both the greatest?

11.3 The total flow rate through a 1 cm radius tube of length 1000 cm is 10 liters per minute. What is the average velocity of the fluid?

11.4 For laminar flow in a tube of radius R, by what factor does the shear stress experienced by a cell increase if it is located at $r = 0.5R$ versus $r = 0.05R$? By what factor does the exposure time to stress increase or decrease when situated closer to the wall versus in the center of the tube?

11.5 Blood packing is a procedure some athletes use to prepare for highly aerobic sports such as marathon running and cross-country skiing. The procedure entails drawing blood and separating the red cells from the plasma months before the event. Just before the event, the red cells are infused back into the athlete in order to increase hematocrit. The theory is that the oxygen-carrying capacity of the athlete's blood will be increased; hence, endurance and performance will be enhanced. Recently, drugs that increase hematocrit have been used instead of the red cell isolation, storage, and infusion method. Apart from the questionable ethics, from the biofluid mechanics viewpoint, are there potential disadvantages to this strategy?

11.6 A 2-liter saline bag is elevated 0.5 meter above a patient in bed. The viscosity and density of saline are 1.2 cP and 1.05 g/cm^3, respectively. Saline infusion helps maintain a patient's fluid and electrolyte balance. The saline flows under the influence of gravity through a sterile, plastic tube of inner radius of 0.20 cm and length, 1 m. To infuse a patient, a 20-gauge hypodermic needle can be used of length 2 cm. The needle is inserted into a vein, which allows the saline to drain into the patient.

(a) What is the flow rate that saline drains from the plastic bag when drainage occurs just through the plastic tube? Assume that the net pressure driving the flow is due to the 0.5 m high column of saline.

(b) What is the drain time for infusion when the needle is used to connect the patient to the saline source, when the same assumption is made?

(c) Why may the actual flow rate be less than that calculated in part (b)?

11.7 A vessel of length L and radius R is replaced by two parallel vessels of equal length, $0.5\ L$, and equal radii, $0.2\ R$. How does the pressure drop of the branched vessels compare to the single vessel alternative?

11.8 Show that if a vessel of length L is replaced by two parallel vessels of length $\frac{1}{2}L$, then $R_B \geq 0.707\ R_U$ must be the case in order for the resulting pressure drop of the branched circuit to not exceed that of the single vessel alternative.

11.9 Convince yourself that the Casson equation actually captures the yield stress, variable apparent viscosity, and then almost Newtonian behavior of blood as strain rate is increased.

(a) Plot τ versus $\dot{\gamma}$ for a typical blood sample where $H = 0.45$, $a\alpha = 1.84$, and $C_F = 0.3$ g fibrinogen/100 ml. On the same curve, show how a Newtonian fluid such as plasma would behave. (Hint: Expand the Casson equation so that τ appears, as opposed to $\tau^{1/2}$.)

(b) Show on the curve three key aspects of blood behavior.

11.10 Instead of branching into two vessels, assume that a vessel of length L_U and radius R_U is replaced by three vessels of length L_B and radius R_B. Derive an equation that predicts how the ratio of pressure drop developed in the three-branch network compares to the single-vessel scenario ($\Delta P_B/\Delta P_U$).

11.11 Starting with the velocity distribution for laminar flow (Equation 11.4b) and $\tau = -\mu\ dv/dr$,

show that the pressure drop depends directly on how large the shear stress is at the wall: $\Delta P = 2\ \tau(r = R)L/R$. (Note: Here a negative sign links τ and dv/dr because as r increases, v decreases and τ increases.)

11.12 If someone's hematocrit was increased from 0.45 to 0.60, estimate how much more power would be expended by the heart in order to circulate 5 liters of blood every minute. Assume that $a\alpha = 1.84$ and $C_F = 0.3$ g fibrinogen/100 ml.

11.13 A heart-lung machine pumps blood from the patient and oxygenates it; thereafter the blood is returned to the patient. The tubing that conveys venous blood to the machine needs to be replaced quickly. There are four sterile tubings available in the OR. Which tubing from the choices below would you use to avoid red cell damage and why? Assume the same volumetric flowrate (Q) is pumped regardless of the tubing used.

(a) $R = 0.5$ cm, $L = 1$ m.
(b) $R = 1$ cm, $L = 1$ m.
(c) $R = 1$ cm, $L = 1.2$ m.
(d) $R = 1.5$ cm, $L = 1.5$ m.

11.14 Which of the choices below will provide the lowest pressure drop needed to drive flow?

(a) Two parallel vessels: $R = 0.2$ cm, $L = 10$ cm.
(b) Two in-series vessels: $R = 0.2$ cm, $L = 10$ cm.
(c) Three parallel vessels: $R = 0.2$ cm, $L = 10$ cm.
(d) Three parallel vessels: $R = 0.15$ cm, $L = 6$ cm.

11.15 A Newtonian fluid is flowing down a sloped, flat surface. Imagine rain draining off a sloped roof. The flow system and spatial dependence of velocity are shown below.

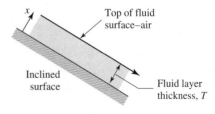

$$v(x) = \frac{A}{\mu}\left[1 - \left(1 - \frac{x}{T}\right)^2\right]$$
v = velocity
A = constant specific to slope angle, etc.
μ = viscosity
x = distance from sloped surface
T = fluid layer thickness

(a) Based on analogies to other cases, will shear stress be the greatest when $x = 0$, 0.25T, 0.5T, 0.75T, or T? Write a sentence that explains your selection.

(b) Using the given velocity dependence on x, prove mathematically that the value of x you chose in part (a) corresponds to where the maximal shear stress occurs.

11.16 Using sketches of how τ depends on $\dot{\gamma}$, contrast two fluids: (1) shear thickening with no yield stress and (2) shear thinning with a finite yield stress.

11.17 An artery with a diseased section is shown below; plaque and other deposits (dark blobs) line the inner wall of the vessel.

Blood exit

An arterial bypass operation is proposed where vessels are surgically removed from another location in the body and then grafted as shown below.

(a) Explain why a patient with clogged arteries exhibits high blood pressure, assuming that blood behaves like a Newtonian fluid. Cite relevant physical/mathematical relationships when explaining.

(b) If the grafts available to the surgeon vary in their radii and lengths, which type (length-radius combination) is best to use for the bypass surgery? Explain your reasoning.

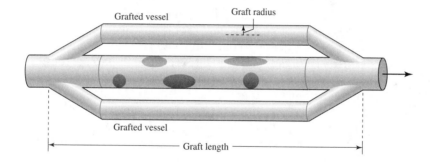

Grafted vessel   Graft radius

Grafted vessel

Graft length

CHAPTER 12

Biomaterials

12.1 | Purpose of This Chapter

The previous chapter discussed how bioengineers develop complex and useful devices such as artificial hearts by using their knowledge of fluid mechanics and an understanding of blood's cellular composition and rheology. Insuring that damage to cells does not create a new problem, while trying to remedy a current medical condition, is a major accomplishment. The choice of materials to use for constructing artificial organs and implants is also an important consideration. Material reliability is essential, because a malfunction can be life-threatening. Moreover, even when a malfunction is not immediately life-threatening, the repair or replacement of an implanted device often requires major surgery, which has risks associated with it. The materials used also must not be toxic or engage the body's immune and wound-healing systems. When such systems are activated, the device's performance may be compromised, or new problems created, such as blood clots and strokes.

Research on biomaterials and technology development covers a lot of ground. Practitioners possess knowledge of material science and engineering, which provides the basis for inventing or selecting materials that will provide adequate strength, durability, and other desirable properties. This knowledge also enables biomaterial engineers to characterize the function-related properties of natural materials in order to learn more about how nature has engineered them to work so well. Apart from advancing our basic understanding, these measurements provide performance targets for engineered replacements. Most biomaterial engineers also understand what responses from the body should be avoided when materials are implanted, and how those responses are triggered. This knowledge also guides their decisions on material design and selection. There are experts on using metal alloys, polymers, and other classes of materials. The **Society for Biomaterials** (http://www.biomaterials.org/pub.htm) is a major international organization that fosters research and the communication of results in journals and at regular meetings.

Parts of the field of biomaterials engineering have recently merged with that of tissue and cell engineering. Interesting work is now being pursued on combining cells and engineered appliances and devices for the purpose of building hybrid artificial organs. Biomaterial knowledge is also integral to constructing the scaffolds used in tissue engineering. Chapter 8 provided one example, the use of degradable polyesters. An implanted system must not engage the body's defense systems, while at the same time, the goal of a functional, replacement tissue must be achieved.

This chapter will first describe three useful concepts in materials engineering to introduce how properties and design goals can be quantitatively assessed and communicated. Then, the body's wound repair system will be surveyed, since it presents a hurdle that a biomaterial engineer must surmount. An overview of the immune system will then be presented. Examples of how the three properties drive use and performance as well as provide the impetus for material modification will conclude this chapter.

12.2   Three Basic Quantifiable Features of Biomaterials

Biomaterials must possess adequate mechanical properties to perform a task such as the repair of weight-bearing joints. The surface of the material is also where materials engineering meets biology. If this interface is not properly chosen and managed, then it is quite conceivable that despite all the other advantageous qualities of a particular material, its use could create more problems than it is solving. Thus, one mechanical property, **elastic modulus**, and two surface-specific properties, **surface roughness** and **wettability**, are presented.

Elastic modulus.   When a material has a tensile force applied to each end, as shown in Figure 12.1, it tends to stretch. The increment in stretching relative to the unstressed length is known as the **strain** (ε), which on a percentage basis is defined as

$$\text{Strain} = \varepsilon = \frac{[\text{Stretched length} - \text{Original length}]}{\text{Original length}} 100. \qquad (12.1)$$

Initially, the strain will be proportional to the **stress** (σ = Force/cross-sectional area) applied as shown in Figure 12.1. The initial ratio of the stress to the strain is called the **elastic modulus** ($E = \sigma/\varepsilon$); this ratio is also referred to as the **Young's modulus**. When units of newtons and meters are used for stress (Force/Area = newtons/m^2), the modulus has units of pascals, because strain is dimensionless. However, because strain can be a small number for a stiff material, modulus values are often reported in giga Pascals (1 GPa = 10^9 Pascals). The greater the value of E, the more resistant to deformation the material is, and thus such a material is "stiff." The acquisition of stress-strain data is called **tensile testing**. Standardized protocols have been developed by the **American Society for Testing and Materials (ASTM)** in order to facilitate the reproducibility and comparison of data from different laboratories. Values for a variety of materials can be found in Table 12.1.

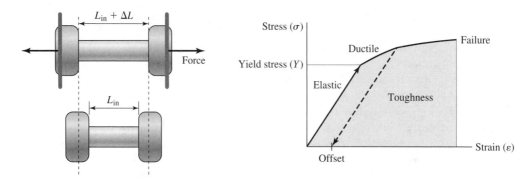

FIGURE 12.1 Tensile testing of a material. (Left) When a force is applied, the material's initial length (L_{in}) is increased to $L_{in} + \Delta L$. (Right) The resulting stress-strain curve and some of its important characteristics.

TABLE 12.1

Examples of Materials and Modulus of Elasticity	
Material	Modulus of Elasticity 10^6 N/m^2
Diamond	1,200,000
Steel	210,000
Copper	124,000
Aluminum	73,000
Glass	70,000
Bone	21,000
Concrete	17,000
Wood	14,000
Plastics	1,400
Rubber	7

If a material returns to its original length (shape) after the stress is removed, it is known as **elastic** under the stress-loading conditions applied. However, after experiencing higher stresses, some permanent deformation may occur as evident by the material relaxing to a length that is distorted from the original. The permanent deformation is known as an **offset** and the condition is shown on the stress-strain curve in Figure 12.1. As the stress is continually increased, eventually the material will fail by breaking into two parts. Prior to failure, the apparent elastic modulus will decrease and the material is easily deformed; then it is called **ductile**. The stress required to break the material specimen is called the **tensile strength**. Because this stress is not divided by a strain value, the numerical value of the elasticity can exceed the magnitude of the tensile strength. The area under the stress-strain curve has the units of work and energy; this area is called the **toughness** and is indicative of how much a material can "take" before failure occurs.

Surface roughness.   Vessels and organs are composed of cells and molecules; hence, biological surfaces have a characteristic texture and a low, molecular-scale value of **roughness**. The simplest way to quantify surface roughness is to obtain a height-profile along some track on the surface by using a sensitive gauge. Typically, an average peak height exists with smaller and larger peaks appearing with some frequency. The irregularity of a surface can be quantified by computing how a series (n) of height measurements at different spatial positions along the track ($s(x)$) compares with the average ($\overline{s(x)}$) height. The value calculated by the following equation (σ_s) is called the **root mean square (rms) roughness**, and the equation should be reminiscent of how the standard deviation on an exam is calculated:

$$\sigma_s = \sqrt{\frac{1}{n}\sum_{x=0}^{x}[s(x) - \overline{s(x)}]^2}. \tag{12.2}$$

Equation 12.2 provides some insights, but it also has limitations. A few large peaks could account for the same value of σ_s as would more numerous smaller peaks. There may also be a regular pattern to the roughness. Consequently, many other parameters have been defined to describe surface roughness and topography. Some parameters attempt to capture whether features exist such as periodicity in the roughness pattern. Different parameters are inspired by statistics and also the mathematics used for **image analysis**. Image analysis involves using computerized, mathematical calculations to process the bits of data that compose an image for the purpose of revealing patterns and/or unique features. The codification of discriminating features in a scan of one's fingerprint by a crime lab is one example of image analysis. The analysis of surfaces through surface profiling and image analysis tools is also important after a biomaterial has been used. Analyses of **explanted devices** (surgically retrieved from a patient after implantation) such as artificial hip joints can shed light on the wear mechanism and other fates that limited their useful life or led to failure.

Surface wetting and contact angle.   The testing of a material or surface treatment with real biological molecules and cells, as will be described later, provides a gauge of how an implant will perform in the human body. However, it is also desirable to relate such performance to another fundamental property of the biomaterial surface. Establishing a relationship will contribute to the theory of biomaterial design. Another pragmatic outcome is that it becomes possible to detect possible alterations in material surface properties *prior to implantation* when variations in material fabrication, implant manufacture, implant storage, and/or sterilization protocols occur.

Another easily measured, yet informative property is the **contact angle** a water droplet makes with a surface. You have probably already encountered the contact angle principle in some aspect of your daily life. For example, you may have noticed how "water beads up" on a freshly waxed car or piece of furniture. Indeed, "beading" is a popular way to display the efficacy of wood surface treatments. Water beads up on such a treated surface because the attractive forces between water molecules exceeds the force between the surface and water. A

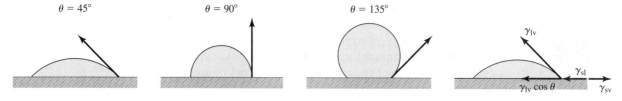

FIGURE 12.2 The contact angle is geometrically defined as the tangent of the interface formed where liquid, solid, and gas phases intersect. From left to right, examples of 45, 90, and 135-degree contact angles are shown. The force balance on the far right dictates that the contact angle formed is the resultant of three interfacial tensions: liquid-vapor (glv), solid-liquid (gsl), and solid-vapor (gsv).

water-attracting (i.e., **hydrophilic**) surface would instead tend to foster the adhesion of water molecules with the result that a droplet would smear out to maximize the area of water-surface contact.

High or low "beading" depends on the observer and is a fuzzy description. A more precise and quantitative description is provided by a contact angle measurement. The geometric definition of a contact angle and a sense of what the angle's magnitude means are shown in Figure 12.2. The contact angle is the angle assumed by a liquid where three phases (liquid, solid, and gas) intersect. An angle less than ninety degrees corresponds to liquid spreading as opposed to beading. In this case, the liquid is referred to as **wetting the solid**. A zero contact angle indicates the extreme of **complete wetting**. When the angle exceeds ninety degrees, we obtain the familiar beading that occurs on water-repelling surfaces. The surface in this case is called **nonwetting**, and the higher the angle, the more extensive the beading that occurs.

The contact angle attained depends on the resolution of three forces. To acquire a sense of what is involved, omit the presence of a solid and consider water in contact with air. The cohesive forces between water molecules are high and greater than the attraction between water and air molecules. Thus, the water molecules tend to maximize their association with each other while minimizing the number of air molecules that they contact. The result is that the area of the surface that divides the water and air phases is minimized. This is why water droplets and mists attain a spherical shape in air as opposed to an ellipsoid or other shape. For a fixed volume, a sphere is the geometric shape that has the minimum surface-to-volume ratio.

The tendency of water molecules to stick together manifests as a **surface tension**. A **tension** is a force per distance. Here, the surface tension force acts along the phase boundary and the distance scale is the along the water drop perimeter. The result is that water and other liquids behave as if they have an elastic skin. Because force per distance is dimensionally equivalent to work (or energy) per area, some people find discussing interfaces easier in terms of **interfacial energies** (energy per area). In other words, when three phases are in contact, there is an energy scale and total energy associated with the areas that divide the gas-liquid, gas-solid, and liquid-solid phases.

The two equivalent views permit one to use mechanical or thermodynamic viewpoints to describe **interfacial phenomena**, which are the events and interactions that occur when materials of different phases encounter each other. To illustrate, consider an insect landing on a water surface. The insect's legs dimple the surface and rather than sinking, the insect's mass is supported akin to a person standing on a trampoline. The insect is supported because the water surface has a mechanical tension. However, work must be performed to distort the interface. Taking instead the energetic viewpoint, the air-water interface has an energy and area associated with it. The dimple creates more area and thus alters the energy from the prior equilibrium state.

The contact angle attained depends on how the various surface tensions balance out. For a liquid in contact with a solid and air, the liquid-vapor, solid-vapor, and solid-liquid surface tensions are denoted as γ_{lv}, γ_{sv}, and γ_{sl}, respectively. Based on the force diagram in Figure 12.2, the contact angle is provided by the **Young Equation:**

$$\gamma_{lv} \cos \theta = \gamma_{sv} - \gamma_{sl}. \qquad (12.3)$$

From inspection, you can find that as the solid-liquid surface tension increases relative to the solid-vapor surface tension, $\cos \theta$ become negative, which indicates that the contact angle exceeds ninety degrees. This force then helps drive the attainment of the nonwetting condition.

The contact angles formed at water-biomaterial interfaces are commonly measured before and after a treatment that is intended to enhance function or **biocompatibility.** A treatment may seek reduced wetting to minimize interactions with biological molecules or cells. However, the binding of proteins and cells to surfaces can involve multiple interactions and structural changes involving both the material surface and the adsorbing entity. Thus, some treatments may attempt to fix electrical charges on the surface in order to "repel" negatively charged cells or molecules. In this case, the surface may become more wetting. In both cases, however, contact angle measurements can indicate whether the surface was indeed modified. Such an assurance is useful because subsequent tests can be time-consuming and expensive, so it is important to determine whether it is worthwhile to proceed.

Contact angle measurements are quite good for detecting whether regions of a surface differ or if variability in manufacturing or storage conditions results in batches with different surface properties. Additionally, environmental influences on materials can be detected. For example, some workers report that the contact angle formed at a polymer-air-water interface can change if the person performing the measurement is wearing strong perfume or cologne. The fragrance molecules can adsorb to the polymer surface and thus alter the surface and its interfacial energetics.

Overall, the basic contact angle measurement or variations are well established in biomaterials engineering. Contact angle measurements provide a convenient way to describe and communicate the nature of a biomaterial surface and whether a given treatment affects wettability. Other applications include quality control, and assessing the effects of aging and the impacts made by storage or implant environment.

12.3 | Body Response to Wounding

Having considered three properties, we now review the biological phenomena that material surfaces will be exposed to, or can help initiate. Thereafter, examples of how the bulk and surface properties of materials play a role in performance will be provided.

Blood coagulation. The five liters of blood in a human are capable of considerably more than circulating oxygen-carrying, red blood cells. Blood can be viewed as a highly engineered tissue that carries a tool kit for fixing problems and the "intelligence" to know when to react *and* when to desist. When blood vessels are damaged or severed, blood reacts to form a patch that isolates the leak and stems blood loss. The process of forming a patch is referred to as **coagulation** or **clotting**. The patch is called either a **clot** or **thrombus**. The clot is composed mostly of a protein, **fibrin**. The fibrin network can entrap red blood cells as shown in Figure 12.3, resulting in further restriction of blood flow and leakage from a wounded area.

Clot formation stops after the immediate area of damage is dealt with, which is fortunate because if the process were not well-controlled, entire vessels would contain clots, or the process would happen haphazardly. The unnecessary blocking

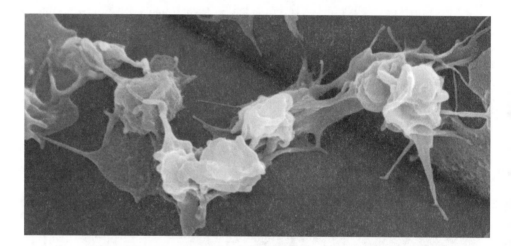

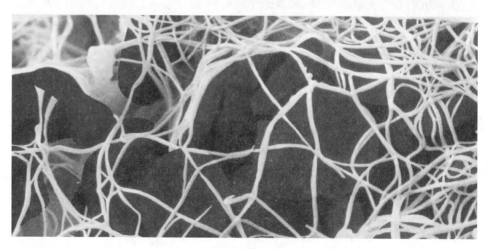

FIGURE 12.3 (Top) Activated platelets and (bottom) a blood clot containing a fibrin network and trapped red blood cells.

of blood supply can lead to **ischaemia** (oxygen deprivation) and tissue damage. These reactive features of blood help to preserve blood inventory when vessels are cut or ruptured. The regulation and preservation of blood volume is called **hemostasis**, a term akin to homeo*stasis*.

When a blood vessel is damaged, three events can occur. First, the radius of blood vessels decreases in an attempt to restrict the blood flow. Second, small platelets aggregate in an attempt slow blood loss. **Platelets** are smaller than red blood cells, disc-shaped, and lack nuclei. Platelets are the product of differentiation of cells in the bone marrow; hence, they are not cells capable of dividing. When they are exposed to the structural protein **collagen** in blood vessels, **platelet activation** occurs. This means that platelets change shape and acquire protrusions that facilitate contact and adhesion. Chemicals stored within platelets are also released to stimulate the next steps in wound response. How platelets appear after activation can be seen in Figure 12.3.

In larger vessels, the third and major event is clot formation. Clotting can occur by two pathways: **intrinsic** and **extrinsic**. The intrinsic pathway uses molecules that are solely present in the blood. The extrinsic pathway uses blood-born factors, plus the activity of some cells that compose the blood vessels. That is, the blood constituents and vessel cells "communicate" to coordinate clot-building activities.

Many molecular factors are involved in either the intrinsic and extrinsic pathways. The key feature of the pathways is that **cascade amplification** occurs. Cascade amplification occurs when one amplified output serves as an input into another amplifier. Several music amplifiers coupled by microphones can illustrate how a cascade amplification system works. Envision that the output from an amplifier connected to a strummed guitar is picked up by a microphone placed in front of the speaker. If the microphone is connected to a second amplifier, the first amplifier's output will be enlarged by the second amplifier. Several serial microphone pickups and amplifications can significantly enlarge the first amplifier's output.

A slimmed-down schematic of how clotting occurs is shown in Figure 12.4. The end-product is an enzyme, **thrombin**, which is derived from an inactive precursor, **prothrombin**. Thrombin catalyzes the conversion of a large protein, **fibrinogen**, to **fibrin**. Fibrin molecules tend to stick together to form the network that can build a clot. Cascade amplification of the clotting process occurs due to the high turnover number of enzymes. An early reaction results in the formation of Enzyme-1 from an inactive precursor. This enzymatic reaction product, in turn, is used to catalyze the next step, which produces another active enzyme. The enzyme-catalyzed turnovers have a multiplying effect. To illustrate, assume that the turnover numbers of the enzymes equal 10 s^{-1} and the time scale is 10 seconds. One molecule of enzyme-1 can produce 100 molecules of enzyme-2 in ten seconds. Over the 10 seconds, enzyme-2 can catalyze the conversion of up to 10,000 molecules of fibrinogen to fibrin. The process is further accelerated by the stimulatory effect of thrombin. The amplification provided by each enzyme output being used as a catalyst for the next reaction, in tandem with the stimulatory effect of thrombin, results in a rapid clotting response.

General biomaterial engineering implications of blood coagulation.

Coagulation can also occur when blood encounters a foreign surface. The clots that

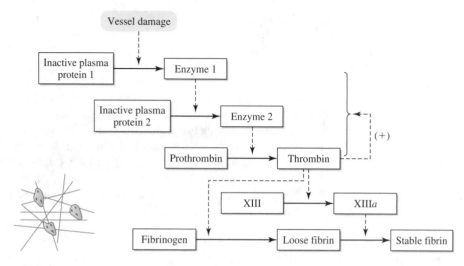

FIGURE 12.4 Simplified schematic of the clotting cascade. The end-product, thrombin, is produced by a sequence of reactions that each produce an active enzyme from an inactive precursor. Each enzyme assists in catalyzing the next step. The compounding of turnover numbers results in signal amplification and rapid clotting response.

form may interfere with the function of the implant. More importantly, the clot material may also detach, resulting in a blockage elsewhere in the circulatory system. A blockage may manifest as a stroke. A clot fragment is called an **embolus** and when detachment and blockage of a vessel occurs, an **embolism** is experienced. A stroke is a common manifestation of an embolism and it involves the restriction of blood flow to the brain.

Embolisms and strokes are major concerns. Indeed, two of the first five recipients of the AbioCor total replacement heart experienced strokes weeks to months after implantation in 2002. Researchers localized the problem to a plastic cage that is involved with fitting the device to an artery. Newer versions of the replacement heart are not fitted with this cage. Apparently, the cage provided a surface and/or source of irritation that triggered the coagulation response of blood. Arterial flows close to the heart can be large and provide substantial shear forces for the detachment of clot material.

Anticoagulation drugs such as **heparin** and **coumadin** can be administered to interfere with the clotting process. Many anticlotting drugs either slow the clotting cascade or stimulate the molecular processes that the body uses to destroy clots by binding to the molecules involved. Other drugs such as **streptokinase** and **tissue plasminogen activator** directly attack and destroy new clots. New clot-dissolving products such as TPA are the result of genetic engineering research and development along the lines covered in Chapter 8.

12.4 | Immune System Defense

The body's immune system provides another layer of defense. This system attacks and neutralizes foreign cells and substances. If an implanted biomaterial engages the immune system, the results can be damage to surrounding tissue and/or compromised implant performance.

The immune system is quite complex, yet there are similarities shared with other systems. Binding reactions are key to initiating the immune response as well as the execution of the defense. Additionally, a series of molecular- and cell-mediated amplification steps occur that result in the rapid onset of a defense in reaction to an insult. Interestingly, there are short-term and long-term responses mounted by the immune system that challenge the biomaterials engineer.

Triggering and Response.   The immune response is triggered when substances that are not native to one's body are recognized. Such substances are called **antigens**. Antigens are typically particular carbohydrate or protein molecules that are associated with foreign entities such as cancer cells, transplanted organs, bacteria, and viruses. Foreign particles and other molecules can also be antigenic. A "piece" of the antigen, as opposed to the whole antigen, is actually responsible for an immune response. Each "piece" of an antigen that is associated with an immune response is called an **antigenic determinant** or alternately, an **epitope**.

To illustrate how recognition and triggering can occur, a cell called a **macrophage** can ingest and degrade particles. The ingestion process is called **phagocytosis**. When a foreign microbe possessing an antigen is ingested by a macrophage, the microbe is likewise degraded. However, degradation is not complete. Rather, some recognizable remnants of the microbe's antigen, in combination with a class of **major histocompatibility complex proteins**, are sent to the macrophage's surface. The surface-bound combination can, in turn, bind to some proteins on the surface of another type of cell in the immune system. The specific binding event plays a role in all the signaling events that activate and direct the immune system response.

The immune defense entails producing **antibodies** and killing infected native cells and "invaders." Antibodies are large proteins that can bind a particular antigen. Most antibody molecules have two antigen binding sites. Antibodies help with vanquishing an invader in a number of ways. For example, antibodies with multiple binding sites can aggregate individual microbes into large particles. Therefore, each time one **phagocyte** ingests and degrades one particle, many microbes are removed from the body. Concerning cell killing, native cells that are infected with a virus can present an alien "signature." This presentation allows **killer cells** in the immune system to recognize them and remove them from the body in an effort to contain an infection. This mechanism is also important in checking for the presence of cancer cells and removing them.

Some small molecules not necessarily associated with viruses and bacteria can also induce an immune response only after they combine with other molecules. A **hapten** is such a small entity. Substances in the background can also heighten the immune response, although it cannot be determined exactly how they participate in the response to a specific antigen. Such substances are termed **adjuvants**.

Overall responses to biomaterial implantation.   When biomaterials are implanted into the body, there is either a **normal injury response** or another, much less desirable response. The responses are mounted against the surgery involved and against the material itself. **Acute inflammation** normally occurs when the biomaterial is nontoxic or otherwise reasonably compatible with the body. Immune system cells first ingest and destroy the damaged tissue and debris associated with the implant. Blood clotting mechanisms such as platelet activation also operate. Thereafter, fibroblasts migrate to the area and synthesize a natural matrix composed of the protein collagen. This matrix fosters the filling in of the wounded area with new cells and tissue. After these responses are complete, the implant will have a **fibrous capsule** around it.

The undesirable alternative is for the acute inflammation response to become a **chronic inflammation**. In this case, the implanted material and/or the damaged area presents a volume of material that cannot be readily cleared by macrophages, and the body is still striving to eliminate the biomaterial. Instead, **foreign body giant cells** appear and accumulate on the scene. The appearance of giant cells is a bad sign, and if the response is serious, tissue and possibly implant damage can result. The only solution now known to deal with a serious, chronic inflammation is to remove the implanted material. Because removal requires another surgery, and all surgical procedures have a certain amount of risk associated with them, biomaterial engineers strive to avoid this complication.

12.5 Examples of the Role of Mechanical Properties of Biomaterials

Elastic modulus. The elastic modulus of skin and bone are roughly 1 and 20 GPa, respectively. Corrosion-resistant metal alloys have much higher modulus values; a typical value for stainless steel is 200 GPa. How the modulus of an implanted material compares to a natural material can be quite important. Sometimes, a much higher modulus for the implanted versus a natural material is desirable. For example, when a metal fixture is used to secure a broken bone, one would like the union to be strongly held together to foster healing. If the fixture was "stretchy," then the breakage area could deflect when a load was applied, which does not foster fast and even healing.

On the other hand, when a degenerated hip joint is replaced by an artificial hip, a metal stem is inserted into the leg bone. The metal stem inside bone can take on some of the mechanical load normally supported by the leg bone. When bone tissues are not mechanically loaded, as can occur during long space flights under reduced gravity conditions, loss of bone mass can occur. Thus, if the implant takes on too much of the stress the bone normally contends with, over the long term bone mass may decrease, which could weaken the union between bone and the implant.

Overall, a mismatch between the modulus of implanted biomaterials and natural materials can be desirable, or introduce side-effects that could lead to trade-offs between desirable and undesirable consequences. These trade-offs can be difficult to predict in advance which accounts, in part, for the evolutionary nature of implanted biomaterials.

Surface roughness. One obvious consequence of surface roughness occurs when the implant has a moving part. If surface roughness is high, then "high spots" called **asperities** can make contact with the opposing surface. If the two surfaces are hard and rough, then the asperity contacts can result in significant friction. If the rough surface is harder than the opposing surface, then the soft surface will experience wear and become marred by scratches or plough tracks. Apart from the friction, surface wear, and reduced lifetime, another important consequence is that **wear**

particles can be shed. The consequence of such particles on the biological response to an implanted material can be high, as will be discussed later in this chapter.

12.6   Examples of Biomaterials Engineering Strategies That Attempt to Minimize Clotting through Surface Modification

The arsenal of established and new drugs reduces the risk of blood clots and provides ways to treat clots when they occur. However, because all drugs have side-effects, biomaterial engineers continually strive to conceive of ways to limit the onset and extent of the clotting process through the design of the material's surface chemistry and morphology. Numerous strategies have been developed. One obvious strategy is to engineer a surface such that cellular and molecular interactions are minimized. Here, the idea is to make the surface as passive as possible, such that platelet adhesion and other clot formation events do not occur. A second and opposite strategy is to encourage cell adhesion and in-growth. In this case, the aim is to encourage alteration of a material's surface after implantation to the point that the cells and molecules presented are those that naturally appear in the body. Lastly, chemical signals can be integrated with the material for the purpose of "jamming" the signals that initiate and propel clot formation. Examples of these different strategies follow.

Limiting roughness.   One obvious strategy is to process biomaterials to achieve a smooth surface. The rougher a surface is, the more surface area exists for molecular and cellular adhesion. Such interactions with blood constituents can initiate coagulation. Therefore, the AbioCor artificial heart designers used smooth plastics to construct the pumping chamber materials and surfaces. The smooth surface in combination with the fluid flow pattern and resultant shear forces are intended to limit the deposition of biological material on the blood-exposed surfaces.

Surfaces that repel cells and molecules.   To discourage protein and cell interaction further, chemical treatment of a biomaterial surface is an option. The adsorption or covalent attachment of polyethylene oxide (PEO) polymers to blood-contacting surfaces is one frequently used chemical treatment. Polymers were introduced in Chapter 8 by using polyglycolic acid as an example of a tissue engineering scaffold (see Figure 8.3). Recall that a polymer is a large, chainlike molecule where repetition of a constituent structure occurs. A physical model of a polymer would be a chain of beads, where each bead is the same color. A **copolymer** is formed from two different monomers; it is comprised of different segments or "blocks" that often differ in chemical properties. Thus, envision a copolymer to be a chain of beads where the beads have different colors, and it is also possible for branches to exist.

Attaching polymers to a surface can discourage protein and platelet adsorption for a number of reasons. One mechanism is called **steric blocking**. Here, the polymer molecule's surface coverage and extension from the surface blocks a protein's or platelet's access to the surface, much like how, in U.S. football, an offensive lineman

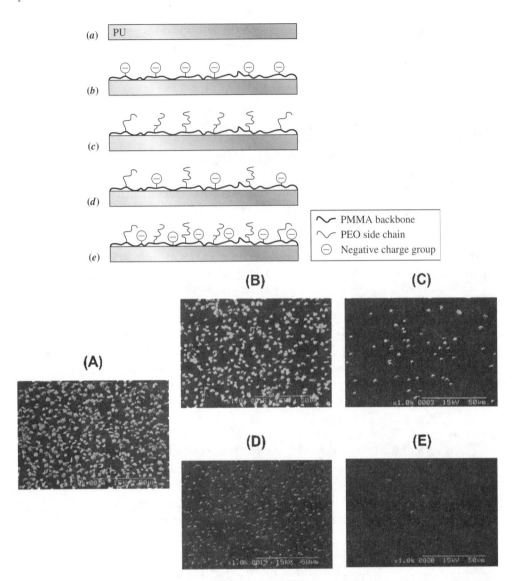

FIGURE 12.5 Outline of a surface engineering strategy and the effects on platelet adhesion. (Top) The raw material is polyurethane (PU). Different surface possibilities are envisioned by coating polymers onto the PU surface: (a) none, (b) hydrophobic polymethyl methacrylate (PMMA) backbone adhered to the surface with negative charges protruding from attached moiety, (c) PMMA backbone and polyethylene oxide chains (PEO) protruding, (d) PMMA backbone anchoring polymer to surface with negative charges and PEO chains protruding, and (e) PMMA backbone anchoring polymer to surface with more negative charges and PEO chains protruding. The corresponding outcomes in terms of platelet adhesion after two hours of exposure of platelets to treated surfaces. Preparations C and E exhibit the least platelet adhesion.

Reproduced with permission from: MMA/MPEOMA/VSA copolymer as a novel blood-compatible material: Effect of PEO and negatively charged side chains on protein adsorption and platelet adhesion. Jin Ho Lee, Se Heang Oh, *Journal of Biomedical Materials Research*, 60, 1, (2002): 44–52 (http://www3.interscience.wiley.com/cgibin/fulltext/89013899/main.html,ftx_abs).

tries to protect a quarterback. Of course, the interaction between PEO and the molecules and cells that are being denied access to the surface must be neutral or repulsive, or the PEO would simply serve as a way to link the surface to proteins and cells.

One demonstration of this technology is shown in Figure 12.5. Copolymers were synthesized of **methyl methacrylate (MMA)**, **polyethylene oxide (PEO)**, and **vinyl sulfonic acid (VSA)** in order to passivate a polyurethane (PU) surface to protein and platelet adsorption. The MMA block was intended to adsorb to the PU. The PEO and VSA blocks were intended to discourage protein and platelet adsorption. VSA was chosen for the negative charge it can provide to its segment of the polymer. Because platelets and many proteins possess net negative charges, VSA may provide forces that repel platelets and proteins.

Contact angles for the raw and treated PU materials were measured. Untreated PU exhibited a water contact angle equal to 57 degrees. The treated materials presented reduced contact angles ranging from 40 to 45 degrees. The surfaces were thus more wetted by water, which is consistent with the hydrophilic nature of the attached polymer chains. In this case, a contact angle measurement confirmed that a chemical modification indeed occurred. Additionally, how the material properties changed made sense based on the knowledge of how the polymers interact with water.

The results were encouraging, as indicated by the diminished extent of platelet adsorption on surfaces bearing the highest PEO content (Figure 12.5; C and E). The surface treated with the highest PEO- and VSA-containing copolymer was particularly effective, as measured by the low platelet adsorption over two hours. Protein adsorption results were interesting as well. Some surfaces displayed low albumin adsorption, while other proteins, such as fibrinogen, were found to avidly bind. Other surfaces displayed low adsorption of all blood proteins tested. Low protein adsorption or preferential adsorption of albumin over fibrinogen is very important. Albumin coating tends to "hide" the surface from the blood coagulation process, whereas surfaces with adsorbed fibrinogen are linked to platelet adsorption.

In general, it is challenging to predict exactly how the adsorption of a particular protein will be influenced by the chemistry of a surface. While proteins possess a net charge, there are patches of different charges on a protein's surface that still may permit an electrostatic interaction to occur between a protein and a surface. Moreover, proteins can unfold or change conformation when they adsorb onto a surface. Changes in a protein's three-dimensional structure can introduce new ways for a protein to interact with a surface.

Active research is underway throughout the world to understand protein- and cell-surface interactions better, so that material passivation methods can be improved further. Additionally, passivation technologies are sought that work over time frames longer than those used in laboratory studies. Increasing the strength of the bonds that retain adsorbed copolymers and using covalent bonds to attach polymers to surfaces are among the avenues being pursued by biomaterial researchers.

Surface treatment to inhibit clotting cascade.   Another approach is to treat a biomaterial surface so that even if the cascade mechanism is activated, blood clotting proceeds to a limited extent. Heparin, as noted earlier, can be administered to minimize clotting. **Heparin** is a sulfonated, polysaccharide chain with a molecular weight that ranges from 3,000 to 30,000. In essence, heparin is a naturally occurring chain of sugar molecules that possess sulfate groups. Heparin is an inhibitor of thrombin. The extent of inhibition increases after it interacts with another molecule present in the blood.

Ways have been devised to use the anticoagulating properties of heparin at the site it is most needed rather than dispersing the molecule throughout the blood stream via an injection. By attaching heparin to a biomaterial surface, the blood in contact with the surface is inhibited from clotting, whereas blood elsewhere in the body "sees" less heparin. The advantage over intravenous administration is that clotting can occur elsewhere in the body in case an injury occurs, while clotting is minimized in the vicinity of the implanted material.

One application is the treatment of catheters. When a catheter is inserted and wiggled through blood vessels, damage can result that activates the blood-clotting cascade. To avoid such an adverse outcome, heparin coating solutions can be used to pretreat catheters and other surfaces before insertion into the body. Often, the heparin is dissolved in an alcohol-water solvent. Alcohol is used to provide some protection against microbial infection, which is another major complication of biomaterial implantation. The material is dipped into the solution and then dried. After implantation, the heparin slowly dissolves and diffuses into the blood, thereby raising the concentration of heparin in the blood proximal to the implanted material. The protection against clotting lasts as long as it takes for the surface-bound heparin to dissolve and diffuse away. Thus, such treatments provide protection from clotting over the time scale of hours to days. A similar approach has been used to treat arterial stents, which are tubes that are inserted into blood vessels after balloon angioplasty to prevent the collapse of an artery after balloon deflation.

Surface design that works with nature.   Another approach is to engineer a material to do the opposite of what the prior two examples strive to achieve. Instead of engineering passivation or clotting inhibition properties into a biomaterial's surface, the different approach is to foster maximal interaction between the biomaterial and biological materials.

The HeartMate LVAD provides an example. The material of the blood-pumping chamber is deliberately designed to be rough and encourage cell and protein adsorption. The intent of using high surface area materials is to rapidly and extensively coat the blood-contacting surfaces with a variety of cells and molecules found in the patient's body. The coating is thought to allow the biomaterial surface to "blend in" and thus not trigger wound repair and blood coagulation responses. Sintered titanium spheres and rough-textured polyurethane are used to provide high surface areas for blood-contacting surfaces. The incidence of thromboembolisms has been encouragingly low when engineered, rough surfaces are used in the HeartMate LVAD.

12.7   Examples of Immune System Links to Biomaterials

Haptens.   Metals are often used in orthopaedic implants and bone repairs due to their strength and workability. Nickel-titanium alloys are amongst the materials employed. Except when corrosion is excessive and ions are released, these implants do not generally elicit a toxic or immune response. However, when used in applications where skin contact can occur, **contact dermatitis** (rash) can result. In this case, Ni^{2+} functions as a hapten when it binds to a protein. The nickel-protein combination activates **Langerhan's cells**, which are cells in the skin that are capable of migrating to lymph nodes and presenting antigens to other cells involved with activating the immune system. Those with nose and other piercings can experience this allergic reaction, as well as people who wear nickel-containing jewelry.

Specific antibody responses. The repair or replacement of large arteries such as the aorta is challenging because it is difficult to harvest comparably strong and large vascular material from elsewhere in the patient. Thus, synthetic materials such as Dacron are used as grafts for replacing sections of blood vessels. Woven and knitted fabrics are available.

Essentially, a "tube" of Dacron fabric is surgically implanted. The blood-contacting surface was initially hoped to be colonized by the **endothelial cells** that line blood vessels. Such colonization would endow the graft with a surface with low reactivity to blood. Colonization was reported in early studies in animal models. However, colonization is not as extensive in humans; it is confined to the ends of the graft that are sutured to a blood vessel. A layer of fibrin is instead established over the bulk of the blood-contacting surface. Consequently, the level of thrombosis is not as low as initially hoped and while manageable, thrombosis can be a post-operative complication.

Graft design presents an interesting optimization problem that illustrates the challenging and interesting nature of biomaterials design and engineering. Fabric porosity is one variable. If a fabric is too porous, the blood loss through the graft's pores immediately after implantation could be life-threatening. On the other hand, if the porosity is too low, then capillary in-growth and cell migration could be impeded. Woven fabrics have lower porosity than knitted fabrics. Thus, knitted fabrics are contacted with the patient's blood prior to implantation and administering heparin. The preoperative coagulation reactions reduce the porosity. An alternative to pretreating graft material with blood is to use knitted grafts that have been infused with collagen.

Graft **compliance** is another variable. Compliance can be defined as the change in vessel diameter that occurs per unit of pressure imposed. As described in the prior chapter, normal vessels are elastic and expand when ventricle contraction occurs. If a graft is not likewise compliant, then blood flow through the graft may be restricted. Additionally, the vessel the graft is attached to will undergo dilation when the circulatory system pressure rises. If the graft's diameter tends not to change in synchrony with the feeder and drainage vessel segments, stresses may be imposed on the sutures that connect the graft to the vessel. The other extreme of excess compliance for a graft is not desirable either. If too compliant, the graft will swell and possibly invade or irritate surrounding tissue. Excessive compliance may also result in permanent shape distortion.

The antigenicity of the polymers used for grafts is yet another variable. Some studies with animals have suggested that polymer-specific antibodies may be generated (Schlosser et al., 2002). These antibodies, if also produced by humans, could provide a clinical index for the short and long term response of the patient to the graft.

Immune responses to implanted materials have to be kept in perspective. The clinical situation prior to implanting an arterial graft is typically serious. If an immune response is kept in check, then the trade-off is positive in that both life span and quality have been elevated. However, documented and proposed immune responses motivate biomaterial engineers to modify or replace materials with synthetic or tissue engineering alternatives.

Adjuvants.   The recent controversy over silicone implants illustrates how a material with many desirable properties might present unanticipated problems after a long time frame of usage. An ongoing debate is underway on whether implanted silicone can function as an adjuvant. Such disputes and the ability to find supporting and refuting data in these cases explain why the biomaterials engineering field tends to cluster around similar materials, and the introduction of alternatives is done cautiously.

As an aside, it is important to note that adjuvants are not all necessarily bad. Many cancer treatments and ongoing clinical trials involve elevating the patient's immune system so that malignant cells can be rapidly found and killed. Experimental implants for vaccine delivery to livestock also exploit adjuvants. The implanted material contains antigen plus factors such as **interleukins** (Lofthouse, S.A., Kajihara, M., Nagahara, S., Nash, A., Barcham, G.J., Sedgmen, B., Brandon, M.R., Sano, A., 2002.) The latter are molecules that the immune system produces in the process of stimulation. The antigen and adjuvants slowly leach out from the material, thereby keeping the immune system continually "on alert" and poised to deal with a bacterial infection that can plague livestock such as sheep.

Wear particles.   Other long-term effects can be presented by implanted materials. Implanted hip joints have improved the quality of life for thousands and the surgery has become almost routine (see Figure 12.6). However, the bearing surfaces,

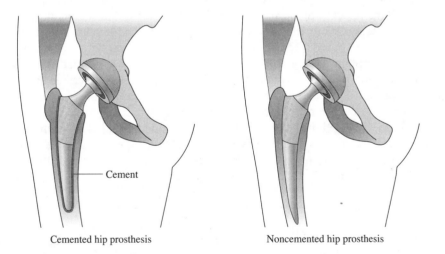

Cemented hip prosthesis   Noncemented hip prosthesis

FIGURE 12.6 Cemented versus noncemented hip implants. The noncemented design extends further into the bone. The rough surface and greater extension is intended to foster more bone in-growth and a tighter and longer-lasting fit. Osteolysis can occur when wear particles stimulate the immune system. The bone loss can cause the implant to loosen, which requires surgical intervention.

which are typically composed of the polymer polyethylene, can wear with use and shed small particles. Other particles and debris can also be generated from the materials used. These small particles are believed to activate macrophages. The immune response results in erosion of the surrounding bone that secures the implant. Bone loss (**osteolysis**) can, in turn, result in the implant loosening. Consequently, the combination of wear and loosening limits the lifetime of hip implants.

There are two designs in use: (1) cemented and (2) noncemented hip implants. One advantage of the cemented design is that the modulus of the cement is less than that of the metal inserted into the bone. Thus, the cement layer can act as a "shock absorber." The noncemented hip implant is more demanding to install, because its larger size requires more fitting and customizing to insure that the gap between bone and the implant channel is not excessive. If the gap exceeds a few millimeters, then bone in-growth will not be sufficient to "grab" onto the implant. The noncemented design was intended to lessen the incidence of cement weakening and bone loss by intensifying bone in-growth and contact. Current work is concerned with using alternative materials for the bearing surfaces, in order to lower the production of wear debris. Other workers study explanted implants to ascertain how the loading and other mechanics of use led to surface wear. More insights from this work may provide leads on how to design implants for an even longer lifetime.

REFERENCES

MMA/MPEOMA/VSA copolymer as a novel blood-compatible material: Effect of PEO and negatively charged side chains on protein adsorption and platelet adhesion. Lee, J.H. and Oh, S.H. *Journal of Biomedical Materials Research*. 2002. 60(1): 44-52.

Immunogenicity of polymeric implants: Long-term antibody response against polyester (Dacron) following the implantation of vascular prostheses into LEW.1A rats. Schlosser, M., Wilhelm, L., Urban, G., Ziegler, B., Ziegler, M., and Zippel, R. *Journal of Biomedical Materials Research*. 2002. 61(3): 450-457.

Injectable silicone implants as vaccine delivery vehicles. Lofthouse S.A., Kajihara M., Nagahara S., Nash A., Barcham G.J., Sedgmen B., Brandon M.R., and Sano, A. *Vaccine*. 2002. 20(13): 1725-32.

EXERCISES

12.1 A rough surface has steps that are 20 μm high and long; the steps repeat every 40 μm. The pattern is shown below. What is the ratio of surface area of the rough-to-perfectly flat surface? Assume that the material has a width (W, the dimension into the page) that exceeds the thickness.

12.2 Height profiles measured from four different surfaces are tabulated below. For all the surfaces, the heights at similar distances along a path of fixed length were measured. The heights are in arbitrary units. How do the values of σ_s compare? Do any of these profiles illustrate the limitation of the σ_s measurement, and

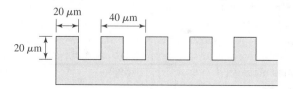

why other parameters have been defined to quantify surface roughness?

Distance	Height Material 1	Height Material 2	Height Material 3	Height Material 4
0	1	0	2	0
1	0	1	0	1
2	1	0	2	1
3	0	1	0	1
4	1	0	2	3
5	0	1	0	0

12.3   What is the difference between homeostasis and hemostasis?

12.4   What does an activated platelet look like and what is one property it acquires after activation?

12.5   Hemophilia is a disease that results when at the genetic level, a mutation has occurred that leads to factor XIII being nonfunctional. What is the consequence to blood clotting when factor XIII is absent?

12.6   How can blood coagulation reactions lead to ischaemia?

12.7   Why could administering too much heparin to a patient recovering from arterial graft surgery prove to be lethal?

12.8   For one cubic centimeter total volume of water, compare the total surface areas when the water assumes a cubic versus a spherical shape. Which shape with the same total volume has the least surface area?

12.9   Teflon is a hydrophobic material. Would the water-air-Teflon contact angle more likely be 45 or 120 degrees?

12.10   A solid material prior to treatment has a water-air-material contact angle equal to 40 degrees. A polymeric material, which will be used to coat the other material, exhibits a water-air-polymer solid contact angle equal to 100 degrees. The coating process involves dissolving the polymer in a solvent. Brushing the solution on the solid material and then allowing for the solvent to evaporate leaves the polymer behind. After the coating process is completed, the contact angle at a water-air-coated solid is measured. Below are

different outcomes. For each outcome, provide an explanation for the observation.

(a) Depending on where on the surface the measurement is made, contact angles are either 40 or 100 degrees.

(b) Everywhere the measurement is performed, the contact angle is 110 degrees.

12.11   What is the difference between an antigen and a hapten?

12.12   A biomaterial implanted years ago in patients has now been found to be an adjuvant. What may be the outcome for the subset of patients who later underwent total hip replacement surgery?

12.13   Would a synthetic polymer be a better material for stitching a wound, as opposed to a thread made from animal-derived protein?

12.14   An artery is grafted with a synthetic material as shown below. If the synthetic material is not compliant, sketch how the vessel-graft combination will appear when pressure peaks in the blood circulatory system.

12.15   A tube is compliant. When the tube is not pressurized, the radius (r_o) is 10^{-2} m. Due to compliance, the radius (r) will increase linearly with pressure at low to moderate pressures, according to $r = r_o(m) + 0.005\Delta P(Pa)$. If water flows through a 0.10 m long segment at a rate 10^{-5} m³/s, how does the pressure drop needed to sustain flow through a totally noncompliant (zero stretch) tube compare to that of the compliant tube?

Pharmacokinetics

C H A P T E R

13

A potential new drug has been discovered. Perhaps, fluorescence microscopy or FACS (presented in Chapter 7) was used to establish the mechanism of action and efficacy of the drug at the single-cell level. If the drug acts by binding to a biomolecule, then perhaps the concentration that needs to be established in the blood stream or within an organ has been estimated based on the ligand binding analyses presented in Chapter 6.

Now to get from a potentially breakthrough drug to one that is routinely prescribed by a physician, dosage and administration protocols must be established. This means that some amount of the drug can either be ingested, injected, or continually infused in order to achieve an effective concentration in the body. There are advantages and disadvantages for each administration route. Moreover, each will result in a different drug concentration-time profile. Ingestion or injection will result in a peak concentration in the body, while infusion will eventually result in a steady level. The questions that now arise are *"How much drug needs to be administered so that the peak level is sufficient for action, but does not prompt excessive side effects?"* or *"What concentration of drug should be infused so that the appropriate plateau level is established?"*

Because an ingested, injected, or infused substance has a long voyage to complete before its target is hit, these are not trivial questions to answer. For example, an ingested drug must first pass through the digestive system into the blood stream. After entering the blood stream, the drug may nonspecifically bind to blood proteins such as albumin. Such binding can also affect how rapidly the drug is eliminated, as suggested by exercises 7 and 8 in Chapter 9. Degradation in the acidic digestion system or within the liver may also occur. The latter processes and others combine to affect the **therapeutic concentration** the target "sees."

Experiments with animals can shed some light on the different fates a drug or other compound may experience. Such tests are, however, expensive, and if excessive, ethical questions can arise. Therefore, if a mathematical model of the body can assist with making dose to final therapeutic level predictions, then less animal testing, less trial-and-error with human subjects, and less expense is required to provide a safe and well-understood therapeutic product.

The science and practice of developing models that describe the fate of drugs and **xenobiotics** (foreign substances from the environment) in the human body is called **pharmacokinetics**. The analysis tools resemble the mass balance and differential equation solving methods introduced in Chapters 2 and 7.

This chapter will provide an introduction to pharmacokinetics. While useful, the mathematical models are highly simplified and abstracted descriptions, and thus have limitations. Consequently, other strategies have recently been proposed. One strategy is to use connected tissue culture reactors in order to replicate the salient aspects of animal physiology. This chapter will conclude by describing one use of cell-based surrogates to supplement pharmacokinetic models and to minimize the use of animal testing.

13.1 | Pharmacokinetic Modeling Basics

Compartment notion. Pharmacokinetic models lump together different organs and systems within the body into connected, functional units over which unsteady-state mass balances can be performed. A functional unit is called a **compartment**. The name is chosen to suggest that both capacitance (recall Chapter 3) and a specific function can be assigned. For example, the stomach and digestive system receive an ingested drug in liquid form. The drug will have a certain initial concentration in the digestive system, which will decrease as it enters the body. Once in the body, the drug can be eliminated by excretion and/or metabolism. In this case, the body can be viewed as one functional unit with inputs and outputs, as shown in Figure 13.1.

Example of model development. Based on Figure 13.1, a mathematical model can now be developed that accounts for the concentration of the drug in the body. It can be assumed that the rate at which the drug leaves the digestive system is proportional to the concentration present in the digestive system. This can be rationalized by assuming that diffusion through the intestinal wall to the rest of the body occurs. The more the concentration in the digestive system exceeds the much lower concentration in the body, the greater the diffusion rate. Likewise, the rate at which the drug is excreted or degraded can be envisioned to increase as the drug concentration increases. Putting these notions together yields

$$\text{Rate drug concentration drops in digestive system} = dC_D/dt = -k_D C_D, \quad (13.1a)$$

$$\text{Rate of drug metabolism} = -k_E C_B, \quad (13.1b)$$

where C_D and C_B refer to the drug concentration in the digestive system and body, respectively. The rate constants and time are denoted, respectively, as k_D, k_E, and t.

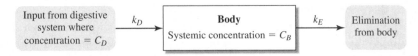

FIGURE 13.1 Basic single compartment depiction to develop a mathematical pharmacokinetic model that predicts the time-dependent level of a drug that enters the body from the digestive system and is then eliminated. The rates at which the drug enters and leaves the body depend on the concentrations C_D and C_B as well as the rate constants, k_D and k_E.

The time-dependent value of drug concentration in the body is the resultant of the inputs and outputs. When the volumes of the digestive system and body are V_D and V_B, the rate at which concentration changes in the body is given by

$$V_B \, dC_B/dt = k_D V_D C_D - k_E V_B C_B. \tag{13.2}$$

Because the volumes are constant over time, by dividing by V_B and defining k_D' as $k_D V_D/V_B$, Equation 13.2, can be written more compactly as

$$dC_B/dt = k_D' C_D - k_E C_B. \tag{13.3}$$

Equation 13.3 has two variables with respect to time, C_B and C_D. If one variable could be eliminated, obtaining a solution would be easier. We desire to predict how C_B changes with time, so the appropriate strategy is to eliminate C_D. From Equation 13.1a, C_D can be eliminated in Equation 13.3 by using

$$C_D = C_{D0} \exp(-k_D t), \tag{13.4}$$

where C_{D0} represents the initial drug concentration in the digestive system immediately after ingestion*. Replacing the variable C_D in Equation 13.3 with the established time dependence, Equation 13.4, provides

$$dC_B/dt = k_D' C_{D0} \exp(-k_D t) - k_E C_B. \tag{13.5}$$

At last, a differential equation in one dependent (C_B) and independent (t) variable is obtained. This equation cannot be solved by the separation of variables method reviewed in Chapter 7. However, another method exists to yield the answer sought. The curious student can see how the following result is obtained by consulting the appendix:

$$C_B = \frac{C_{D0} k_D' [\exp(-k_E t) - \exp(-k_D t)]}{k_D - k_E}. \tag{13.6}$$

Checking the solution and the behavior indicated. The solution may look complicated, but it can be easily screened to see if it is capable of capturing the dynamics of the drug within the body. Initially (i.e., limit $t \rightarrow 0$), the concentration of drug in the body should be zero, because it is all contained within the digestive system and has not had time to pass into the body. When $t \rightarrow 0$, Equation 13.6 indeed reports that $C_B \rightarrow 0$. After a very long period of time has elapsed, all the drug should have entered the body and then have been excreted or fully eliminated in another fashion. Again, letting $t \rightarrow$ infinity in Equation 13.6 results in $C_B \rightarrow 0$.

Because the drug concentration is zero initially, and again after a long time has elapsed, there must be a maximum in drug concentration at some point between the extreme limits of time (t^{MAX}). The maximum in drug concentration (C_B^{MAX}) will occur when $dC_B/dt = 0$. From Equation 13.6 one can easily determine that

$$C_B^{\text{MAX}} = \left[\frac{C_{D0} V_D}{V_B} \right] (k_D/k_E)^{\frac{k_E}{k_E - k_D}}. \tag{13.7a}$$

$$t^{\text{MAX}} = \frac{\ln(k_D/k_E)}{k_D - k_E}. \tag{13.7b}$$

*Starting with Eq. 13.1a and reviewing the origin of Eq. 2.8 may help to follow this step if it is troublesome.

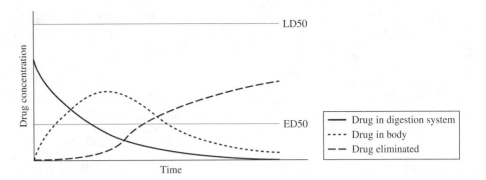

FIGURE 13.2 General time profile for the concentration of a drug that enters the body from the digestive system and is then eliminated by excretion and/or degradation, based on the compartment model shown in Figure 13.1. A peak in concentration occurs. Managing the peak level and the disappearance time is vital to administering the correct dose to a patient. In this case, the drug concentration is in the therapeutic zone for a significant amount of time while remaining well under the toxicity level during the entire time.

The peak level (i.e., C_B^{MAX}) proves to increase as the dose (i.e., C_{D0}) increases and it decreases as the rate of degradation (i.e., value of k_E) increases. The smaller that k_E is, compared to a fixed value of k_D, the more time that elapses before the peak in concentration occurs. The general time-dependent behavior is summarized in Figure 13.2. However, bear in mind that the peak magnitude, time, and shape will depend on the magnitudes and relative values of k_D and k_E.

Does such behavior really occur? Blood alcohol concentration following the ingestion of beer, wine, or liquor tends to qualitatively follow the pattern shown in Figure 13.2, as data shown in Figure 13.3 illustrate. After ingestion, alcohol tends to be rapidly absorbed (i.e., k_D is large). Thereafter, metabolism by liver and other enzymes tends to be slow (i.e., k_E is small). You were introduced to one such enzyme, alcohol dehydrogenase, in Chapter 4. The magnitude of the peak is proportional to the number of drinks ingested, which is in accordance with Equation 13.7a. The time at which the peak occurs, while generally short compared to the time scale for clearing alcohol from the bloodstream, appears to depend on dose, which is not predicted by Equation 13.7b. Because only one degradation mechanism with simple kinetics was assumed, relaxing that assumption could capture this detail.

General utility of solutions to pharmacokinetic models. The general utility of pharmacokinetic models becomes more evident when the quantitative prediction of the time-dependent drug concentration is contrasted to other pharmacological parameters. The dose that results in death for 50 percent of patients is known or can be estimated from animal studies. This dose is typically called the **median lethal dose** **(LD50)**. The smaller dose that results in a therapeutic effect in 50 percent of the population is also known or can be estimated. This other dose is termed as the

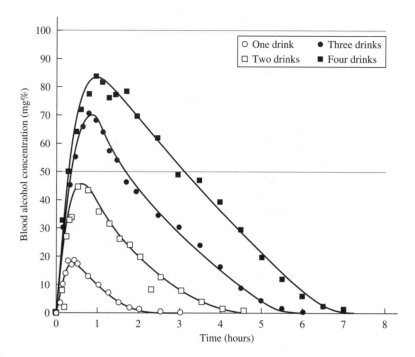

FIGURE 13.3 How blood alcohol concentration varies with time and the number of drinks ingested by eight fasting male subjects. A standard drink is defined as 12 ounces of beer, 5 ounces of wine, or 1.5 ounces of 80-proof distilled spirits. 100 mg% is the legal level of intoxication in most U.S. states. 50 mg% is the level at which deterioration of driving skills begins. The time course of blood alcohol concentration resembles that in Figure 13.2, which was generated by a simple pharmacokinetic model. (Adapted from http://www.niaaa.nih.gov/publications/aa35.htm.)

median effective dose (ED50). The ratio of LD50 to ED50 is called the **therapeutic index**. An index value close to one indicates that a low margin of safety exists.

Now recall Figure 13.2. If the time-dependent drug concentration approaches the value of LD50, a potentially lethal situation exists. The closer the therapeutic index is to unity, the more important it becomes to manage the maximum in the drug concentration. Alternately, for a drug with a large therapeutic index, if the peak in concentration barely reaches the value of ED50, this means that most of the time the drug concentration is below the effective level and a marginal effect may result.

Overall, predicting the time course of drug concentration allows for dosage protocols to be optimized. Additionally, the advantages and disadvantages of different administration routes can be clearly distinguished. Modeling results may suggest that the oral administration of a particular drug is risky, and thus a method such as infusion is more appropriate. Indeed, infusion is commonly used to administer chemotherapy drugs because many have low therapeutic index values.

13.2   Limits of Pharmacokinetic Models and Gaining More Predictive Power

While reading the above example of pharmacokinetic model development, you might wonder how the complex human body can be represented by one compartment with one input and output. Indeed, you may have been surprised to see that a simple model could capture a significant portion of what occurs in a real case, such as the time-dependent variation of blood alcohol concentration (Figures 13.2 and 13.3). Such concern is appropriate, because bioengineers and others must balance model complexity against what they seek to predict.

Simple models are useful because they contain few parameters and are easy to solve. For example, the simple model presented in Equation 13.3 only required estimates of the time scales (i.e., $1/k_D$, $1/k_E$) for two processes, drug release from the digestive system and elimination from the body. A model that was based on *more* mechanistic details could potentially predict how the number of drinks affects the time at which blood alcohol concentration peaks. One possible detail is to account for the kinetics of alcohol oxidation in the liver. Equations akin to those covered in Chapter 6 could be used. Additional compartments that account for the function of the liver and other organs could also be added. These additional details would require the introduction of more parameters whose values, in turn, would have to be obtained from the medical literature or estimated. Each new parameter and model detail contributes to the ability to predict more complex behavior. However, additional uncertainty is also added because each value assigned to a parameter has an estimation or measurement error associated with it. Therefore, the analyst contends with trade-offs between increasing the predictive powers versus adding new uncertainties to a model.

Although the models used in practice contain more compartments and rate processes than the example that led to Equation 13.6, pharmacokinetic modelers tend to develop the simplest model possible based on what prediction is sought. They realize that the model is a highly abstracted representation of the fate of a drug or toxic compound in a human. However, because the average human mind can *simultaneously* keep track of and connect only three different phenomena, these models prove useful because they allow for one to see the time-dependent consequences of three or more processes that interact with each together. One major use of the models is to perform "experiments" on the computer or on paper. For example, one can use the simple model that yielded Equations 13.6 and 13.7a,b to ask and answer questions such as the following:

> *"What happens when drug absorption is much faster than elimination (i.e., $k_D \gg k_E$)?"*
> *"What happens if the drug molecule is altered to be metabolized slower (e.g., k_E decreases ten-fold)?"*

In addition to judiciously increasing model complexity, some new tools for pharmacokinetic analysis have been recently proposed that aim to bolster predictive capability while diminishing animal and human testing. One example entails building a tissue culture model that allows for tissue interactions to be studied without the use of animal test subjects. Separate cultivation systems house cells that comprise different organs. These systems are connected much like organs and tissues in the body are connected by the blood circulation. A mathematical model can also be built to enable the processing of the data obtained from the experimental system.

One such **cell culture analog (CCA)** was prototyped by Sweeney and colleagues (1996) and pursued further by Quick and Shuler (1999). The CCA contained rodent hepatoma (liver) cells and lung cells in order to study whether circulated naphthalene (an environmental pollutant) has a toxic effect on rodent lung cells, and if so, what the mechanism is. Increasing the number of cells and/or inducing certain enzyme activities in the liver compartment increased lung cell mortality, which indicates that liver metabolism can convent a naphthalene to a form that is potentially toxic to other organs. Overall, CCA and other cell-based strategies may find increased use for testing, and for providing the information needed to build "right-sized" pharamokinetic models that can capture key physiological and metabolic phenomena.

13.3 Appendix: Solution of Pharamacokinetic Model

Equation 13.5 corresponds to a more general representation:

$$dC_B/dt + k_E C_B = k_D'C_{D0} \exp(-k_D t) \rightarrow dy/dx + p(x)y = g(x),$$

where $p(x)$ and $g(x)$ are functions of the independent variable or constants. In this case, p and g correspond to the constant and function k_E and $k_D'C_{D0} \exp(-k_D t)$, respectively. Equations that correspond to the general form are called **first-order, linear differential equations (FOLDEs)**. They are **first order** because the *first* derivative of the independent variable is the highest derivative that is present. The descriptor "linear" is used because the independent variable (i.e., y in the general form) appears as the first power instead of as y^2 or a higher power.

FOLDEs can be solved by using an integrating factor, $\mu(x)$, which is given as follows:

$$\mu(x) = \exp\left[\int p(x)\, dx\right].$$

In this case, μ equals

$$\mu(x) = \exp\left[\int p(x)\, dx\right] \rightarrow \mu = \exp\left[\int k_E\, dt\right] = \exp(k_E t).$$

Using μ to solve for the independent variable entails the following steps:

$$y = 1/\mu(x)\left[\int \mu(x)g(x)\,dx + c\right] \rightarrow C_B$$

$$= \exp(-k_E t)\left[\int \exp(k_E t)\,k_D'C_{D0}\exp(-k_D t)\,dt + c\right]$$

$$= k_D'C_{D0}\exp(-k_E t)\left[\int \exp\left[(k_E - k_D)t\right]dt + c\right]$$

$$= k_D'C_{D0}\exp(-k_E t)\frac{[\exp[(k_E - k_D)t] + c]}{K_E - K_D}.$$

The integration constant c can be eliminated with the initial condition that $C_B = 0$ when $t = 0$, which requires that $c = -1/(k_E - k_D)$. Substitution for c provides the particular result that appeared in the text as Equation 13.6.

REFERENCES

A preliminary physiologically based pharmacokinetic model for naphthalene and naphthalene oxide in mice and rats. Sweeney, L.M., Shuler, M.L., Quick, D.J., and Babish, J.G. *Ann. Biomed. Eng.* 1996. 24(2): 305–20.

Use of *in vitro* data for construction of a physiologically based pharmacokinetic model for naphthalene in rats and mice to probe species differences. Quick, D.J. and Shuler, M.L. *Biotechnol Prog.* 1999. 15(3): 540–55.

EXERCISES

13.1 The properties of two drugs are given in the table that follows. Which one may be safer and best to start treatment with?

Drug	LD50 (mg drug /kg patient)	ED50 (mg drug /kg patient)
Mellonzac	0.05	0.02
Burgin	0.1	0.08

13.2 What conditions must be satisfied for two different drugs to exhibit similar peak times when they are orally administered and the model in the text is obeyed?

13.3 What two conditions must be satisfied for two different drugs to exhibit similar peak concentrations when they are orally administered and the model in the text is obeyed?

13.4 Two different drugs and their absorption and elimination properties are given in the table that follows.

By what factor will one drug peak faster than the other when oral administration is used?

Drug	k_D (h^{-1})	k_E (h^{-1})
Mellonzac	10	1
Burgin	100	10

13.5 The infusion of a drug, which provides a drug at a constant rate, is used instead of oral administration. The drug is eliminated from the body at a rate that is proportional to concentration.

(a) Construct a picture of a compartment-based conceptual model from which a mathematical pharmacokinetic model can be developed.

(b) Show that the concentration in the body is given by

$$C_B = \frac{R_{I_{[1-\exp(-k_E t)]}}}{k_E V_B},$$

where C_B, R_I, k_E, t, and V_B refer to drug concentration in the body, infusion rate (moles/time), elimination rate constant, time, and body volume, respectively.

(c) Sketch how the time profile of drug concentration compares to the case when oral administration is used.

(d) Blood measurements indicate that the concentration of an infused drug in a patient is half of the ED50 value. What should be done to bring the concentration up to the ED50 value?

(e) A drug is infused to attain a level equal to 0.1 of its LD50. Now a chemically modified version of this drug has become available. The new version is no longer degraded by the liver, but it can still be excreted by the renal system. Consequently, the effective value of k_E for the new drug is lowered by a factor of ten. The LD50 of the new drug is about the same as the old drug. Based on the

pharamcokinetic model for infusion, what could happen to the patient if the new drug is infused at the same rate as the old drug?

13.6 You can easily solve $dy/dx = -2y$ by using the separation of variables method presented earlier in Chapter 7. However, the problem also represents a very simple first-order differential equation (FOLDE), which can be solved to yield $y(x)$ by using the general method presented in the Appendix.

(a) Use separation of variables to solve for y as a function of x for the case, $y = 1$ when $x = 0$.

(b) When using a new method, it is helpful to first use it on a simple problem that you know the answer to. To gain practice at using the general solution method, see if you can get the same answer as in part (a) when you use the general solution method, which entails first obtaining an integrating factor.

CHAPTER 14

Noninvasive Sensing and Signal Processing

14.1 | Purpose of This Chapter

Imagine being able to look inside a person to spot problems before they become too advanced to treat or to perform a rapid diagnosis when someone is injured. Remarkably, that capability is something we take for granted now because we have all seen ultrasound, magnetic resonance, and other images in magazines or on television. The path to acquiring this capability, however, has been long and challenging. The first hurdle was to overcome the taboo on dissection and anatomical probing of the human body as was established by Popes such as Gregory IX in the Middle Ages. DaVinci, Vesalius, and other figures of the Renaissance broke away from that mindset and instituted the study of anatomy and anatomical-based reasoning. These events captivated artists of the time, as Figure 14.1 illustrates, and inspired a new generation of educators, as the text cover shown in Figure 14.2 attests. From the technical standpoint, this stride provided the basic information and framework with which to pose diagnostic questions and to interpret the results of tests.

Learning how to harness radiating energy and its interaction with biological structures and molecules followed. Indeed, some of the developments and inspirations came from other fields. The infusion of the results into medical science and bioengineering came afterwards. One example is the use of sound to probe for hidden structures in media that do not permit visual examination because they are opaque. For example, the Yugoslavian/Croatian seismologist **Andrija Mohorovicic** (1856–1936) used reflecting sound waves to find a discontinuity in the surface of the Earth. That boundary we now recognize as where the Earth's crust meets the mantle. The reflection of sound off of structures with different densities is the basis of ultrasound technology, which is commonly used

FIGURE 14.1 Invasive dissections were not condoned through the Middle Ages. This image from around 1300 A.D. shows a surgeon, who has removed some tissue, being addressed by a monk and a physician. Some tension between the individuals is evident.

for monitoring fetal development and diagnosing the state of the gall bladder and other organs (see Figure 14.3).

   This chapter will illustrate the science and engineering involved with using radiant energy to probe biological systems. There are two aspects that need to be understood in all applications: (1) how radiant energy interacts with materials and (2) how to extract information from the signal that permits a "picture" to be constructed that has meaning to a clinician. This chapter provides an introduction to these two aspects through the examples of magnetic resonance spectroscopy and imaging. To provide the working fundamentals, we will first discuss how **microwave energy** interacts with biological matter. Then, how a mathematical tool such as the **Fourier transform** enables us to interpret the information that the energy-matter interaction provides will be presented. This chapter concludes with examples of how magnetic resonance spectroscopy and imaging can be used for research and clinical practice.

FIGURE 14.2 A page from Vesalius's *De Fabrica Corporis Humani* (1543) shows how teaching anatomy was avidly embraced at Padua and elsewhere. Students are clearly engaged and pleased to attend class. From the *The Smithsonian Book of Books* by Michael Olmert (Smithsonian Institution Press, ISBN 0-89599-030-X (1992).)

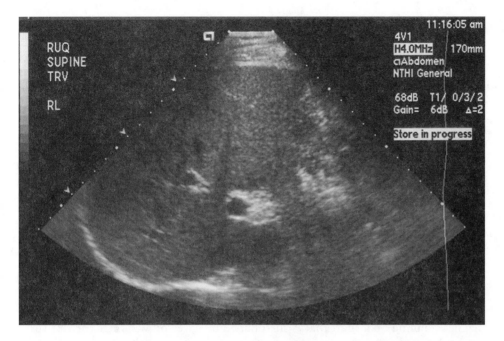

FIGURE 14.3 Ultrasound image of the author's gall bladder. The sound waves reflect differently off tissues and organs with different density. The difference in reflected intensity can be used to create an image.

14.2 | Physics of Nuclear Magnetic Resonance

Nuclear spins and magnetic fields. Some atoms possess a property called **nuclear spin**. This means that they behave as if the nucleus is a charged entity in motion. To see the significance of this property, think back to an early experiment you may have performed or observed. A wire carrying a current can be used to build an electromagnet. By analogy, the apparent spinning charged nature of some atomic nuclei results in the nucleus behaving as if it were a little magnet.

Among the atoms that possess the property of spin are phosphorous and hydrogen. This is significant because biological systems contain substantial amounts of hydrogen- and phosphorus-containing molecules, such as water (H_2O) and ATP, respectively. Thus, it would be potentially beneficial to use hydrogen- and phosphoros-containing molecules inside the body as reporters of functional status.

When a sample of molecules containing atoms that possess the spin property is placed in a strong, static magnetic field, there is a tendency for the "little magnets" to align with the field. This is akin to a compass needle (small magnet) aligning with the magnetic field lines of the Earth and pointing north. The stronger the fixed field, the more spins that will align.

There is an additional feature to the interaction that is a consequence of the angular momentum property of a spin system. In addition to the increased tendency of the little magnets to align with the magnetic field, the angular momentum vectors of the little magnets also rotate about the fixed field direction. This rotation is termed **precession** and it is a resultant of a rotating system possessing angular momentum while being subjected to a torque. To envision this further, recall the experience of a spinning bicycle wheel on a table. The wheel's axle is aligned with the direction gravity acts, which is perpendicular to the table's surface. When the wheel rotates (akin to a nucleus possessing "spin"), an angular momentum vector is aligned along the axle. Gravity also exerts a torque on the wheel. Consequently, the axle winds up rotating in a cone centered about the direction gravity acts (akin to the atom's magnetic "pole" rotating about the direction of the static field). In nuclear magnetic resonance applications, the precessional frequency is called the **Larmor frequency**.

The picture we have so far is that some magnets line up and precess about the fixed field axis, as shown in Figure 14.4. How can this be useful? To see the utility, it is helpful to recall an elementary school experiment that involved moving a bar magnet to and fro in a wire coil. The result was that an electrical voltage and current were induced by the moving magnetic field. This is the basis for how a magnet spinning on a shaft that is surrounded by wire coil windings (or vice versa since relative motion of a magnet to a coil is needed) acts as a generator of an alternating current and voltage. Such a device is in your car: the alternator is used to recharge the battery after the battery becomes drained by powering the starter motor.

So imagine now that we can somehow flick a switch to "ON" and "tip" the resultant rotating magnetization into the horizontal plane, as shown in Figure 14.4. Flicking the switch to "OFF" will result in the magnetization returning to where it was beforehand. The return process involves a moving magnetic field akin to that moving bar magnet experiment. Therefore, if we locate a coil around the sample of nuclei, then the coil will "pick up" the return process in the form of an oscillating voltage that decays with time. The voltage decays because as the magnetization returns to where it was initially, the magnitute of the magnetization in the x-y plane decreases. This time-dependent voltage is called a **free induction decay**.

Interaction of radio frequency energy and spin systems.   How can the spinning resultant magnetization be tipped so that a voltage signal can be picked up, and why is this useful? The energy is provided by switching a radio frequency transmitter "ON" and then "OFF". The actual frequency used provides radiation with the right amount of energy to tip the magnetization. Energy is required to tip the magnetization because the little magnets prefer to line up with the static field. To be aligned otherwise requires energy input to allow the work of "tipping" to be performed. Thus, the frequency is particular to the type of nuclei and the strength of the static magnetic field. The frequency at which tipping occurs also equals the Larmor frequency. Tipping at a particular frequency is known as the **resonance condition** and sometimes referred to as **exciting the spins**.

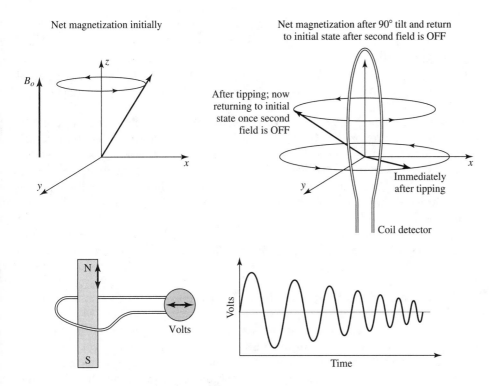

Net magnetization initially

Net magnetization after 90° tilt and return
to initial state after second field is OFF

After tipping; now
returning to initial
state once second
field is OFF

Immediately
after tipping

Coil detector

Volts

FIGURE 14.4 Sequence of events in nuclear magnetic resonance (NMR). In NMR, a coil is used to transmit an rf burst to a sample/subject to cause the spins to tip. The coil then functions as a receiver to detect the voltage produced by the moving magnetic field. (Top, left) The static field tends to align the spins along the direction of the static magnetic field (B_o), which is directed in the z-direction. The net magnetization vector of the aligned spins actually rotates about the z-axis. (Top, right) When briefly exposed to another field that is perpendicular to B_o, the spins tip to the xy-plane. The spins continue to rotate when tipped due to precession. The spins relax back to rotating around the z-axis after the second field is turned off. (Bottom, left) The motion of the net magnetization of the spins enclosed by a coil is akin to a moving bar magnet in a coil. When moving field lines cut the coil loop, a time-dependent voltage and current is induced. (Bottom, right) An oscillating voltage is produced that decays with time each time a spin alignment and tipping cycle is executed.

The physics of magnetic resonance is now well-developed. Here, we summarize the discussion so far, noting some relationships that will be important for understanding spectroscopy and imaging application. We denote the static magnetic field and **Larmor frequency** (or **resonant frequency**) as B_o and ω_o. A typical frequency is 100 MHz. Then,

1. The fraction of the total nuclear magnetizations aligned with the direction of B_o increases with the strength of B_o.
2. The frequency for resonance depends on the nuclei type and increases with the strength of B_o.

3. The signal provided by one free induction decay increases as the number of aligned nuclear magnetizations increases. Thus, the signal increases with the strength of B_o.

4. Adding free induction decays results in a measurement signal; hence, the total signal obtained is proportional to the number of summed free induction decays.

It is evident that the strength of the static magnetic field is very important. Much engineering effort has been invested into building superconducting magnets that provide very strong fields. To provide a sense of the strength, if you were standing 2 to 3 feet away from a typical magnet used for imaging patients while holding a flashlight or other magnetic object, at a minimum the flashlight would be ripped from your hand. You could also be pulled to the magnet if your grip on the light was strong. Indeed controlling the type of materials used in diagnostic environments is a very important safety consideration for clinical personnel and patients. Many tables and other fixtures are made of plastic or other high strength, nonmagnetic material to avoid accidents and injuries.

Shielding and the excitation trick.   Let us return to the little nuclear magnets such as hydrogen and phosphorous for a moment. These elements are found in many molecules that exist in living systems. Each hydrogen (H) or phosphorous (P) atom in a molecule can have different neighbors such as carbon, nitrogen, and oxygen. Because shared and somewhat mobile electrons constitute the bonds between elements in a compound, each nucleus can be in a different environment. This results in varying degrees of **shielding** from B_o. The effect of shielding is to lower the apparent value of B_o that a nucleus "sees." According to the second of the four relationships cited earlier, the consequence is that the different H nuclei, for example, in a molecule will have different resonant (Larmor) frequencies.

The shielding-based spread in excitation frequencies raises the question *"How can a transmitter operate at one frequency tip all the H magnetizations in a sample when they may all differ in excitation frequency?"* The answer lies in an important aspect of statistics. When the result of a Gallup Poll is given on some public opinion question, information on sample size is always provided. Sample size is important because as the size decreases, the answer we obtain is more variable.

To see how variability relates to sample size, consider the thought experiment in the insert found at the end of this chapter. The experiment entails placing you on the roadside for the purpose of counting the number of cars that pass per minute. If a car passes every minute, and there is a one-minute gap between passages, your answer will depend on the time interval (i.e., sample size) over which you count cars. If you hit the stopwatch just as a car passes, count that car, and wait one more minute, another car will enter your view. You will thus score two cars over that time window. If instead, you hit the stopwatch when no cars are in view, you will at most count one car over a one-minute time window. Overall, the variability in your measurement is one car. Now if you made your observation period longer, you would count more cars. There would still be variation in the measurement, but it would become smaller compared to the total cars counted as the observation time increased.

While variability in sampling can be a problem to a pollster, the variability property is cleverly exploited in the operation of the radio frequency transmitter. The transmitter is turned on to precipitate the free induction decay. However, the "ON" time is short and comparable to the cycle time of the radio frequency wave, which consists of peaks and troughs. Thus, there is an apparent variation in the number of peaks that pass during the brief "ON" time. This variation then results in an apparent spread in excitation frequencies. The spread is sufficient to achieve resonance with all the nuclear magnetizations despite their different shieldings. It is quite a clever way to use one piece of equipment, and illustrates the utility of basic statistics.

14.3 Signal Processing: Converting Raw Signal into Useful Information

Now that we see how it is possible to excite all the nuclei of interest, the final aspect to consider is how the signal in the free induction decay can be made useful. Because all the H nuclei and their particular molecular environments, for example, will have different abundances and frequencies, their free induction decays will vary in strength (depending on abundance) and frequency (depending on shielding). Thus, rather than seeing one waning wave, the composite signal received will be the sum of waves of different frequencies and amplitudes as the example with the fictional molecule Carnegionate illustrates (see inset).

This signal, while rich in information, is quite complicated, and it would be useful to sort out all the bits of information buried in the signal. For example, determining what the composite frequencies and amplitudes are would tell us how many different H atoms are present and in what abundance. This is useful information, for example, because each molecule can differ in its nuclear magnetic resonance "fingerprint" because of the different shielding it provides to the constituent nuclei that possess spin. Thus, comparing what we extract from the composite free induction decay to catalogued fingerprints would allow us to identify what molecules are present. Considering that we are obtaining this information from an intact system going about its business, that is an exciting possibility, because we can watch biomolecular events and chemistry in action rather than destructively sampling the system and obtaining such information after the fact.

To process the free induction data to extract the frequency-amplitude components, another mathematical trick is used. This trick is called a **Fourier transform**, which is a key tool used by bioengineers and others in signal processing. The idea behind the transform can be appreciated by performing some computer-aided calculation "experiments." As shown in Figure 14.5a, you can start by graphing a cosine wave, $\cos \omega_s t$, where ω_s and t are a signal frequency and time, respectively. Now on the same graph, plot the product $\cos \omega_s t \cos \omega_t t$, where ω_s and ω_t are the signal frequency and a "test" frequency, respectively. An example is shown in Figure 14.5b, where the signal and the test frequencies are 2 and 3 hertz (Hz) respectively. Now compute the area under the curve formed by the product of the two cosine functions. Assign the area elements above the x-axis a positive value and those below the axis

FIGURE 14.5 The basis of a Fourier transform. (a) Two cosine waves are first plotted: a signal (2 Hz) and a test signal (3 Hz). Multiplication of the two waves provides Figure (b). When the frequencies of the signal and test waves are unequal, the net area under the curve of the multiplied waves is zero as indicated by areas such as *A* and *B* (and others) having negative counterparts, $-A$ and $-B$.

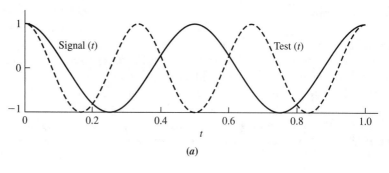

(a)

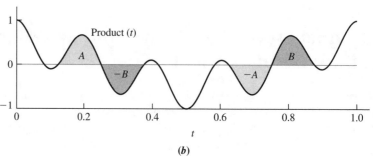

(b)

a negative value. By inspection of the example in Figure 14.5b, you should be able to convince yourself that for every positive area (e.g., *A*, *B*), a corresponding negative one (e.g., $-A$, $-B$) exists.

What you have just done is computed the area under the curve formed by multiplying the signal (2 Hz) and test signal (3 Hz), which is expressed below as

$$\int_0^t \text{Product}(t)\, dt, \tag{14.1}$$

which can be shown to be zero when the signal and the test frequency are unequal. If one instead had *N* discrete signal data points, such as voltage values at different times, the equivalent calculation for when the measurements are made at equal to Δt time units apart is

$$\sum_{i=1}^{N} \text{Product}\,(t_i)\Delta t \tag{14.2}$$

The result again equals zero when the signal and other frequency are unequal. If one "experiments" further using different test frequencies or uses a mathematical proof, it is revealed that when a periodic signal is multiplied by a "test" signal, the resulting area under the product curve is zero unless the signal and test signal have the same frequency.

This can be extended to signals such as free induction decays that possess more than one frequency, and thus are more complicated than a cosine function. Effectively, the raw signal is multiplied by a series of "test" signals that differ in frequency. Thus, sweeping through a range of test frequencies allows one to ferret out the frequency components in the composite free inductions. The result is a frequency spectrum. For example, a free induction decay composed of two decaying cosine waves of frequency 1 and 2 Hz is shown in Figure 14.6a. Using a series of test frequencies, the frequency spectrum produced from the raw signal data yields the spectrum shown in Figure 14.6b. In essence, what the Fourier transform does is to help one determine what notes are in a chord and the strength of each note.

Recall that the frequency of a free induction decay signal depends on the type of nuclei responsible (H, P, etc.), the fixed field strength, and the nuclei's bonded environment (e.g., local shielding extent). Fourier transforming the raw free induction decay data will reveal what frequency components exist. These components, in turn, will thus inform us what nuclei and molecular environments are present in the sample to produce such a signal. This information allows one to identify what molecules are present in a sample. Additionally, their participation in biochemical reactions can potentially be monitored.

The extraction and utilization of signals from intact living systems is one major area of work for bioengineers with a substantial background in electrical and computing engineering fundamentals. Dealing with signals that differ in strength, and recasting the raw information in a way that allows for use and interpretation by clinicians and other practitioners, are major challenges, and a fine record of success exists.

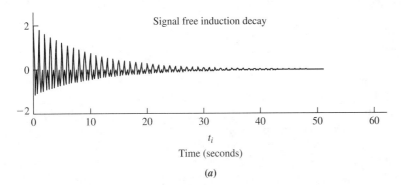

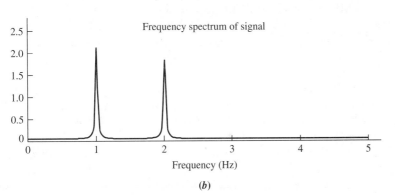

FIGURE 14.6 Illustration of what a Fourier transform yields. The composite free induction decay signal in (a) has 1- and 2-hertz components. Using a series of test frequencies yields the frequency spectrum shown in (b). There are 1- and 2-hertz signal components that differ in amplitude.

14.4 | Nuclear Magnetic Resonance Applications

Basic analysis. A basic use can now be illustrated. A sample contains phosphate and it is desired to measure the phosphorous content. First, the phosphorous nuclei are allowed to align with the static magnetic field. Then the radio frequency transmitter is turned on to tip the phosphorous magnetization. Turning off the transmitter now allows for the detection of a free induction decay. The voltage induced and detected will be proportional to the abundance of phosphorous nuclei. The signal from one free induction decay is generally quite weak. To overcome this problem, the alignment and burst procedure is repeated many times. The free induction decays are summed until they add up to a substantial value. Thus, for a set of free induction decays, we have a summed signal that will reflect the abundance of phosphate. Comparing the measurement to a calibration with a known solution can provide a reasonable estimate of the actual concentration in the sample.

Watching biochemistry in action in living cells. The above was a useful example and it tied together some of the fundamentals, but it did not fully harness nor demonstrate the power of a method such as nuclear magnetic resonance. Ultrasound and other methods that use radiating energy to probe a system have great utility because they enable the interrogation of the structural and chemical interior of a sample without dismantling or destroying the system. This so-called **noninvasive** feature is a great asset for medical research and diagnosis.

An important research application is the observation of chemical reactions occurring in live, *intact* cells. One example is the response of yeast to a potential inhibitor of one of the metabolic reactions that degrades glucose. The time sequence of ^{31}P NMR spectra in Figure 14.7 shows that the effect of the inhibitor on the cells is significant. Impairing the ability to degrade glucose results in decreased ATP concentration within the cells; hence, the cells are energetically compromised. The assertion that the inhibitor blocks glucose degradation is also supported by the data from the cells. Sugar phosphates (recall Figure 3.4), which are the types of molecules that are present upstream of the blocked enzyme, also increase with time. This outcome makes sense because when an enzyme is inhibited, the upstream metabolites cannot be degraded further, so they tend to "pile up" like cars on a freeway when an accident is ahead.

Using NMR to watch how intact, live cells behave over time has increased our understanding of the inner workings of cells. Also evident from the example is that the tool can be quite useful in exploring the mechanism and effectiveness of how a drug may affect the inner workings of cells.

Magnetic resonance imaging. One of the more useful clinical applications of NMR is magnetic resonance imaging (MRI). This method produces the interesting pictures of what lies within the body without having to perform surgery or tissue dissection. **Paul Lauterber** was one of the pioneers in developing the method and early equipment. His work was performed at the **Mellon Institute, Pittsburgh, PA** in the 1960s.

Understanding how MRI works is enabled by recalling item 2 in the summary of NMR physics. For a given nuclei, the resonant frequency increases as the

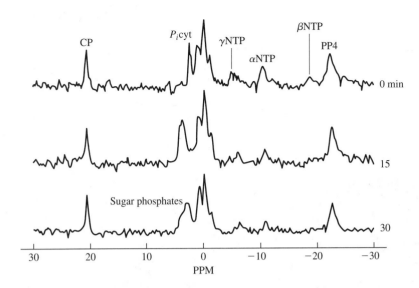

FIGURE 14.7 Response of intact yeast cells to a compound that inhibits glucose degradation (recall Figure 3.4). The time series of ^{31}P NMR spectra reveal that ATP (in NTP signals) declines. Among other changes are that sugar phosphate molecules upstream of the inhibited enzyme reaction accumulate. Author's research, which appeared in *Biotechnology Progress*, 1999, 15:65–73.

aligning field (B_o) strength increases. Thus, if one places a specimen in a magnetic field that is engineered to be stronger on the left than the right, then the nuclei on the left will emit signals at a higher frequency than those on the right. Consequently, each spatial location in an object can be probed with a different frequency. The signal at a given frequency provides nuclei abundance and other information at a particular location.

The example of a box-shaped specimen is shown in Figure 14.8a. The difference between the highest- and lower-frequency component in the Fourier transformed signal will be equal to the field gradient dB_o/dx times the specimen dimension when the gradient is linear. At each point along the field, the same number of nuclei are "seen" in each cross section penetrated by the field lines. Thus, the frequency spectrum has the shape of a simple box, because each frequency is possessed by an equal number of nuclei.

A cylindrical-shaped object provides a more interesting example. As shown in Figure 14.8b, the frequency spread will again depend on the dimension of the object, which in this case is the diameter of the solid. The height of each frequency component will again reflect how many nuclei are present that emit a signal at a given frequency. As the cylinder is pierced by the field lines from right to left, an increasing number of nuclei are encountered until the center of the object is reached. Thereafter, the number of nuclei wanes until they vanish when the left edge is passed. The frequency spectrum thus has a different shape for the cylinder than for the box-shaped object.

By using multiple field gradients and the information from the spectra, the image of a more complicated object such a human head can be assembled as shown in Figure 14.8c. Considerable engineering and analysis goes into building and controlling the magnetic field gradients and radio frequency pulses that allow the information to be generated. Extensive analysis and computation is also needed to process the information in order to detect, for example, the presence of a brain tumor. Such information is useful for diagnosis, and the spatial information plays a key role in the surgical planning process. Human factors and ergonomics have also entered in the

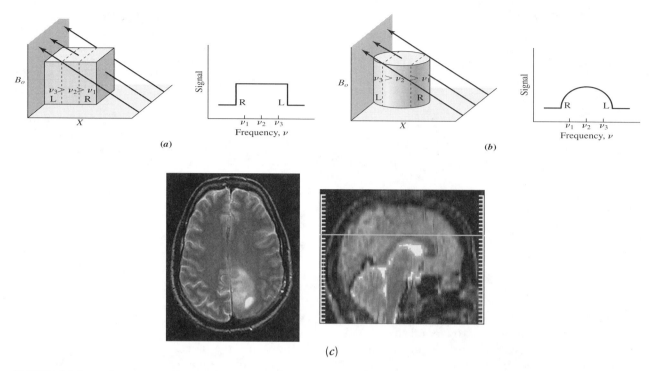

FIGURE 14.8 (a, b) Essence of how NMR imaging is accomplished with field gradients and (c) a clinical example showing the detection of a brain tumor. Clinical images reproduced with permission of *The Whole Brain Atlas* © Keith A. Johnson and J. Alex Becker, all rights reserved. For more information on *The Whole Brain Atlas* ©, see http:// www.med.harvard. edu/AANLIB/cases/case31/mr1-tc1/019-html.

engineering and design of MRI systems. The original design shown in Figure 14.9 used large electromagnet tunnels, which worked well, but made some patients feel confined and anxious about the procedure. Newer designs, as show in Figure 14.9, use an open magnet to lessen the anxiety associated with the imaging procedure.

FIGURE 14.9 A modern open-bore MRI system (left) versus a tunnel-style imaging system (right). Open bore systems have been developed because some patients have experienced anxiety when confined in a tunnel-style imager.

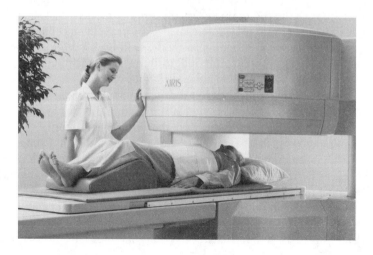

INSETS

A Thought Experiment to Illustrate How a Short Time Creates a Spread in Apparent Frequency of an Event

Your goal is to measure the traffic flow on a rural, bumpy back road. Assume it takes one minute for a car to pass your observation point. For simplicity, assume that the cars are spaced one minute apart. If you count cars for 5 minutes, you can get different answers.

Car

Observation point

Car Counting Experiment Results	
Watch for 5 Minutes	
0–1 No car (N)	Car Passes (C)
1–2 C	N
2–3 N	C
3–4 C	N
4–5 N	C
0.4 Cars/min	0.6 Cars/min

If the sampling time decreases to, for example, 3 minutes, the spread in the answers would grow. If, instead of cars, the peaks or troughs of a wave were counted over a short time, the apparent spread in frequency would likewise increase as the sampling time decreased. Thus, in NMR *short* bursts of radio frequency energy at a fixed frequency are usually sufficient to affect all the nuclei, even though their frequencies can differ.

Signals from Molecules Combine and Need to Be Untangled into Component Contributions

The Carnegionate molecule is shown below. It contains two atoms that excite and emit signals at two slightly different frequencies due to shielding effects.

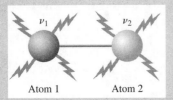

Atom 1   Atom 2

The signals from atoms 1 and 2 will both be decaying waves that differ in frequency.

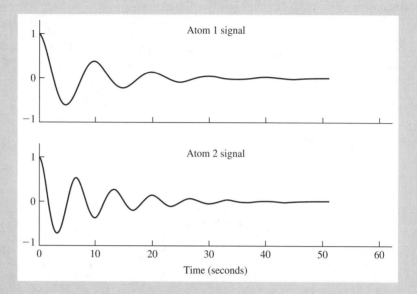

The observed signal will, in turn, be the sum of the signals from atoms 1 and 2.

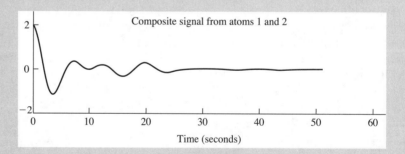

Finding component signals with frequencies associated with atoms 1 and 2 would be desirable, because it would allow us to conclude that the particular molecule, Carnegionate, is present. Moreover, the amplitudes of the component signals will help us figure out how much Carnegionate is present in the sample. The sorting process amounts to working backwards from the bottom composite to the individual signals; sorting can be achieved by using a **Fourier transformation** of the raw signal.

ADDITIONAL READING

Nuclear Magnetic Resonance and Its Applications to Living Systems. David G. Gadian. Clarendon Press. ASIN: 0198546270 (May 1985). (Very well written and concise).

Biomedical Signal Processing and Signal Modeling. Eugene N. Bruce. John Wiley & Sons; 1st edition. ISBN: 0471345407 (November 20, 2000).

Signals and Systems in Biomedical Engineering: Signal Processing and Physiological Systems Modeling (Topics in Biomedical Engineering International Book Series). Edited by Devasahayam. Plenum Publishing Corp. ISBN: 0306463911 (July 2000).

Principles of Magnetic Resonance Imaging; A Signal Processing Perspective. Zhi-Pei Liang and Paul C. Lauterbur. IEEE Press Series in Biomedical Engineering Edited by Metin Akay. IEEE Press, New York, NY. ISBN: 0-8194-3516-3 (2000).

EXERCISES

14.1 The percentage of nuclear spins that are aligned with the static magnetic field in a NMR spectrometer is 1 percent.

(a) What will happen to the percentage aligned and resonant frequency of the nuclei if a spectrometer with a stronger magnet is instead used?

(b) For either spectrometer, do you think that the "ON" time required for the second field to tip the magnetization will increase or decrease when the strength of the second field is increased, and what is your reasoning?

14.2 Based on the questionable presence of lines in the Fourier transformed data, the signal from 100 free induction decays obtained from a cell suspension is weak. What are three possible ways to solve this problem so that the effect of a drug on cell metabolism can be studied?

14.3 Will a high-shielded or low-shielded H nuclei exhibit a lower than average Larmor frequency?

14.4 In a spectroscopy application of magnetic resonance where you seek to watch the chemistry occur within intact cells, which is better, a strong static magnetic field or one that varies within the sample?

14.5 A faucet drips at a rate of 1 drop every 2 seconds. What is the range of dripping rate that will be measured if you count drops over a 5 second observation period?

14.6 Figure 14.5(b) is reproduced below. Indicate how the areas other than A, $-A$, B, and $-B$ cancel.

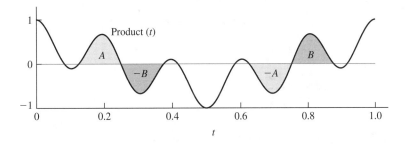

14.7 Below is a complex signal. It contains two waves of frequency 3 and 10 Hz. The component signals have equal amplitudes. If you did not know the component signals and had only the raw data to work with (i.e., below), explain how you would analyze the data to obtain the frequency spectrum (i.e., figure out what the constituent frequencies are). Use words and equation(s) to explain. What would a graph of intensity vs. frequency look like?

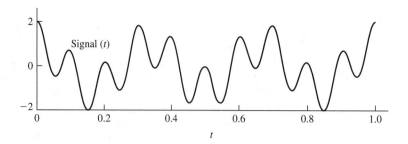

14.8 Below is a spherical object situated in a magnetic field gradient. There is an anomaly in the object, which does not provide a signal. Plot what the "image" would look like on an intensity versus frequency plot (on right).

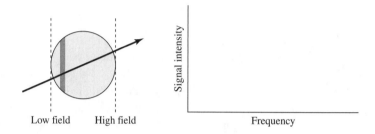

14.9 Below is an object that is subjected to one-dimensional magnetic resonance imaging. The dark region produces less signal than the light region due to a difference in nuclei concentration. What was the field gradient used to provide the "image" (signal vs. frequency plot) shown? Draw your answer on your sketch of the object and be sure to note where the field is high and low.

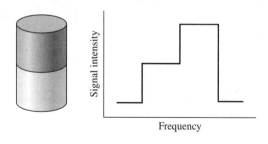

14.10 Prove that the integral which follows is zero over the interval (0, 1 second) if frequency-1 (ν_1, in Hz) does not equal frequency-2 (ν_2, in Hz). Assume that ν_1 and ν_2 are integers. (Hint: use the trignometric identity $\cos(A)\cos(B) = \frac{1}{2}[\cos(A - B) + \cos(A + B)]$.)

$$\int_0^1 \cos(\nu_1 2\pi t) \cos(\nu_2 2\pi t)\, dt = 0.$$

14.11 An object and an MRI image in one plane are shown below. The object possesses a subvolume (dark region) that emits zero signal. Indicate on the object where the static field (B_o) is low and high (i.e., show the (B_o) field gradient).

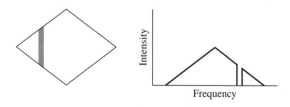

14.12 It is desired to obtain an NMR spectrum of a biological tissue sample to determine the abundance of metabolites. So a spectrum is desired where the signal from each ^{31}P nuclei associated with a given molecule is sought. Despite the best efforts of engineers, the B_o field is not completely uniform. Therefore, when the sample is put into a magnetic field, it is spun as shown below. Explain why the sample is spun and the positive effects on the spectrum.

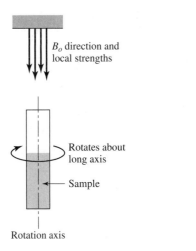

14.13 An inverted "C" chord is played on a piano. Each note is equal in strength. The notes of the inverted chord are G (low note), C (middle note), and E (high note). (The interval between G and C is greater than C and E.) Sketch (i) what the raw signal looks like (roughly is OK) and (ii) what the Fourier transformed data look like. *Label the axes on the two sketches.*

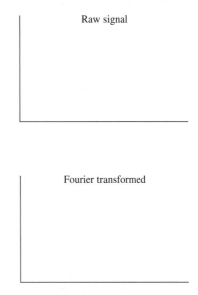

Index